Wave Propagation and Radiation in Gyrotropic and Anisotropic Media

Abdullah Eroglu

Wave Propagation and Radiation in Gyrotropic and Anisotropic Media

Springer

Abdullah Eroglu
Indiana University-Purdue University
Fort Wayne, IN
USA
eroglua@ipfw.edu

ISBN 978-1-4419-6023-8 e-ISBN 978-1-4419-6024-5
DOI 10.1007/978-1-4419-6024-5
Springer New York Heidelberg Dordrecht London

Library of Congress Control Number: 2010933849

Cover design:

Printed on acid-free paper

Springer is part of Springer Science+Business Media (www.springer.com)

Dedicated to my wife G. Dilek

Preface

As technology matures, communication system operation regions shift from microwave and millimeter ranges to sub-millimeter ranges. However, device performance at very high frequencies suffers drastically from the material deficiencies. As a result, engineers and scientists are relentlessly in search for the new types of materials, and composites which will meet the device performance requirements and not present any deficiencies due to material electrical and magnetic properties.

Anisotropic and gyrotropic materials are the class of the materials which are very important in the development high performance microwave devices and new types composite layered structures. As a result, it is a need to understand the wave propagation and radiation characteristics of these materials to be able to realize them in practice. This book is intended to provide engineers and scientists the required skill set to design high frequency devices using anisotropic, and gyrotropic materials by providing them the theoretical background which is blended with the real world engineering application examples. It is the author's hope that this book will help to fill the gap in the area of applied electromagnetics for the design of microwave and millimeter wave devices using new types of materials.

Each chapter in the book is designed to give the theory first on the subject and solidify it with application examples given in the last chapter. The application examples for the radiation problems are given at the end of Chap. 5 and Chap. 6 for anisotropic and gyrotropic materials, respectively, after the theory section. The application examples presented in the last chapter also present the comparison of the device performance using isotropic, anisotropic, and gyrotropic materials. This will help to see how material properties impact the device operation for the specific application presented.

The scope of each chapter in the book can be summarized as follows. Chapter 1 introduces Maxwell's equations and details dyadic analysis, k-domain techniques for general anisotropic media. The derivation of the general dispersion and constitutive relations, and detailed analysis of wave propagation and the dispersion characteristics for uniaxially anisotropic, biaxially anisotropic, and gyrotorpic media are given in Chap. 2 and Chap. 3, respectively. Dyadic Green's function

(DGF) method for general anisotropic media is discussed and DGFs for unbounded and layered anisotropic and gyrotropic media are derived in Chap. 4. DGFs derived in Chap. 4 find an application in Chap. 5 and Chap. 6. In Chaps. 5 and 6, the radiation characteristics of the anisotropic and gyrotropic media are presented. Chapter 7 discusses the wave theory of the layered composite structures. In the final chapter, practical real word engineering application examples are given using isotropic, anisotropic, and gyrotropic materials.

Fort Wayne, IN

Abdullah Eroglu

Acknowledgements

I would to recognize my colleagues at ENI Products, MKS Instruments and Syracuse University for the discussions that helped significantly my research and improvement of the material presented in this book. Special thanks go to my editor, Steven Elliot, for his support during the course of the preparation and publication of this book.

It would not be possible to complete this book without dedication shown by my wife and children. I would like to thank them for their patience during the preparation of this book.

Contents

Chapter 1
Introduction

In this chapter, we will briefly discuss the evolution for the need of the novel materials, what has been done so far, and what needs to happen in the future. We will give Maxwell's equations for general anisotropic media and discuss dyadic techniques that can be used in the analysis of anisotropic medium. We will introduce k-domain and detail how to use in the analysis of the general anisotropic media.

1.1 History of Novel Materials

The ever expanding communication needs have led to the utilization of increasingly higher frequency bands. Systems in use utilize the millimeter and optical bands. Thus, much emphasis is placed today on the development of devices which will perform, at high frequencies, parallel functions to those already available at lower frequencies. Among these is the class of nonreciprocal components.

Using general anisotropic material instead of isotropic material introduces more parameters due to tensorial behavior of permeability and permittivity. Furthermore, an application of a static magnetic field can be used to dynamically control the material parameters of gyrotropic materials. This will bring new degrees of freedom by introducing more parameters to the design of the device. More specifically, materials that are anisotropic and exhibit nonreciprocal behavior can be combined with isotropic, reciprocal substrates in planar circuitry [1]. For example, magnetically gyrotropic or gyromagnetic materials such as ferrites can be integrated with semiconductor substrates such as GaAs or Si to produce nonreciprocal antenna components which are merged with microwave integrated circuit (MIC) structures [2]. Ferrites also have been widely used as the key elements in microwave devices such as phase shifters, isolators and circulators [3–6].

The effect of non-reciprocity has also been observed in electrically gyrotropic or gyroelectric materials such as cold plasma or some of the semiconductor material under static magnetic field. For instance, a 35 GHz isolator using coaxial solid state

A. Eroglu, *Wave Propagation and Radiation in Gyrotropic and Anisotropic Media*,
DOI 10.1007/978-1-4419-6024-5_1, © Springer Science+Business Media, LLC 2010

plasma is proposed by Mcleod and May [7]. In their work, an InSb semiconductor at 77 K (mobility $= 48 \frac{m^2}{V.\ s}$, conductivity $= 1.5 * 10^3 \frac{s}{m}$) with a length $\ell = 28.9\,\text{mm}$ exhibits a 2 dB insertion loss and 30 dB isolation when the frequency is 35 GHz and the magnetic field is 0.2 T. A design of a broadband slot-fed gallium arsenide (GaAs) circulator operating at 77 K from 50 GHz up to 125 GHz is reported by Sloan et al. [8]. Mok and Davis [9] analyzed GaAs phase shifters and isolators at millimetric and sub-millimetric wavelengths.

There is a relentless research on the materials such as gyrotropic and anisotropic to obtain better performing microwave devices. The trend is to use composite structures involving isotropic and anisotropic materials to meet with stringent criteria of military and industrial applications. As a result, it is a need to have the required knowledge to be able analyze structures with anisotropic, gyrotropic and composite structures. This book is intended to provide the knowledge to perform the analysis of microwave devices using these materials.

1.2 Maxwell's Equations

Maxwell equations in differential forms for general anisotropic medium in the presence of an impressed magnetic current density $\bar{M}$ and an electric current density $\bar{J}$ can be written as

$$\nabla \times \bar{E} = -\frac{\partial \bar{B}}{\partial t} - \bar{M}(\text{Faraday's law}) \tag{1.1}$$

$$\nabla \times \bar{H} = \frac{\partial \bar{D}}{\partial t} + \bar{J}(\text{Ampere's law}) \tag{1.2}$$

$$\nabla \cdot \bar{B} = 0\,(\text{Gauss' law for magnetic field}) \tag{1.3}$$

$$\nabla \cdot \bar{D} = \rho\,(\text{Gauss' law for electric field}) \tag{1.4}$$

where $\bar{E}$ is electrical field intensity vector in volts/meter (V/m)
$\bar{H}$ is magnetic field intensity vector in amperes/meter (A/m)
$\bar{J}$ is electric current density in amperes/meter^2 (A/m^2)
$\bar{M}$ is magnetic current density in amperes/meter^2 (A/m^2)
$\bar{B}$ is magnetic flux density in webers/meter^2 (W/m^2)
$\bar{D}$ is electric flux density in coulombs/meter^2 (C/m^2)
ρ is electric charge density in coulombs/meter^3 (C/m^3)

The continuity equation is derived by taking the divergence of (1.2) and using (1.4) as

$$\nabla \cdot (\nabla \times \bar{H}) = \nabla \cdot \left(\frac{\partial \bar{D}}{\partial t} + \bar{J}\right) \tag{1.5}$$

Since $\nabla \cdot (\nabla \times \bar{H}) = 0$, then

$$\nabla \cdot \bar{J} = -\frac{\partial \rho}{\partial t} \tag{1.6}$$

Equation (1.6) also represents the fundamental law of physics which is known as conservation of an electric charge.

In isotropic medium, the material properties do not depend on the direction of the field vectors. In other words, electric field vector is in parallel with electric flux density and magnetic field vector is in parallel with magnetic flux density.

$$\bar{D} = \varepsilon \bar{E} \tag{1.7}$$

$$\bar{B} = \mu \bar{H} \tag{1.8}$$

ε is the permittivity of the medium and represents its electrical properties and μ is the permeability of the medium and represents its magnetic properties. They are both scalar in the existence of an isotropic medium.

However, when the medium is anisotropic this is no longer the case. The electrical and magnetic properties of the medium depend on the direction of the field vectors. Electric and magnetic field vectors are not in parallel with electric and magnetic flux. So, the constitutive relations get the following forms for anisotropic medium.

$$\bar{D} = \varepsilon_o \bar{\bar{\varepsilon}} \cdot \bar{E} \tag{1.9}$$

$$\bar{B} = \mu_o \bar{\bar{\mu}} \cdot \bar{H} \tag{1.10}$$

Permittivity and permeability of anisotropic medium are now *tensors*. They are expressed as

$$\bar{\bar{\varepsilon}} = \begin{bmatrix} \varepsilon_{11} & \varepsilon_{12} & \varepsilon_{13} \\ \varepsilon_{21} & \varepsilon_{22} & \varepsilon_{23} \\ \varepsilon_{31} & \varepsilon_{32} & \varepsilon_{33} \end{bmatrix} \tag{1.11}$$

and

$$\bar{\bar{\mu}} = \begin{bmatrix} \mu_{11} & \mu_{12} & \mu_{13} \\ \mu_{21} & \mu_{22} & \mu_{23} \\ \mu_{31} & \mu_{32} & \mu_{33} \end{bmatrix} \tag{1.12}$$

Then, (1.9) and (1.10) takes the following form in rectangular coordinate system.

$$D_x(B_x) = \varepsilon_o(\mu_o)\left[\varepsilon_{11}(\mu_{11})E_x(H_x) + \varepsilon_{12}(\mu_{12})E_y(H_y) + \varepsilon_{13}(\mu_{13})E_z(H_z)\right] \tag{1.13}$$

$$D_y(B_y) = \varepsilon_o(\mu_o)\left[\varepsilon_{21}(\mu_{21})E_x(H_x) + \varepsilon_{22}(\mu_{22})E_y(H_y) + \varepsilon_{23}(\mu_{23})E_z(H_z)\right] \quad (1.14)$$

$$D_z(B_z) = \varepsilon_o(\mu_o)\left[\varepsilon_{31}(\mu_{31})E_x(H_x) + \varepsilon_{32}(\mu_{32})E_y(H_y) + \varepsilon_{33}(\mu_{33})E_z(H_z)\right] \quad (1.15)$$

1.3 Boundary Conditions

Maxwell equations given in (1.1)–(1.4) accurately represent the electromagnetic fields in a region when the material properties such as μ, ε, or σ are continuous and do not change. If the material's properties vary across some surface, the field vectors are expected to be changed accordingly. It then becomes necessary to establish *boundary conditions* at the interfaces on field vectors when there is a discontinuity and a change in the material properties. This way Maxwell equations give unique and valid solutions everywhere for electromagnetic fields even there is discontinuity at the interfaces. The boundary conditions can be obtained using the integral form of Maxwell's equations. Integral form of Maxwell equations can be written using Stoke's and divergence theorems. Using Stoke's theorem it can be shown that the integral of the curl of a vector over surface area S is equal to the line integral of the same vector around the boundary as shown in (1.16) and illustrated in Fig. 1.1.

$$\int_S (\nabla \times \bar{K}).\hat{n}ds = \oint_L \bar{K} \cdot \hat{l}dl \quad (1.16)$$

When it is applied to (1.1) and (1.2) in the absence of magnetic source current, we obtain

$$\oint_L \bar{E} \cdot \hat{l}dl = -\frac{\partial}{\partial t}\int_S \bar{B} \cdot \hat{n}ds \quad (1.17)$$

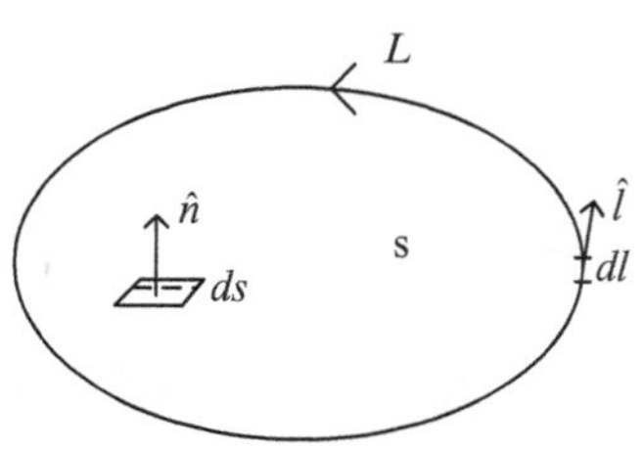

Fig. 1.1 Illustration of the Stoke's theorem

$$\oint_L \bar{H} \cdot \hat{l}dl = I + \frac{\partial}{\partial t} \int_S \bar{D} \cdot \hat{n}ds \tag{1.18}$$

where I is the current and defined by

$$I = \int_S \bar{J} \cdot \hat{n}ds \tag{1.19}$$

We can take the derivative outside of the integral because the surface area does not change with time. Divergence theorem states that the integral of the divergence of a vector over volume V is equal to the integral of the same vector over surface area S which encloses volume V. This is shown by (1.20) and illustrated in Fig. 1.2.

$$\int_V \nabla \cdot \bar{K}dv = \oint_S \bar{K} \cdot \hat{n}ds \tag{1.20}$$

When this is applied to (1.3) and (1.4) in the absence of magnetic source current, we obtain

$$\oint_S \bar{D} \cdot \hat{n}ds = \int_V \rho dv \tag{1.21}$$

$$\oint_S \bar{B} \cdot \hat{n}ds = 0 \tag{1.22}$$

At this point we can apply the integral form of Maxwell equations given by (1.17), (1.18), (1.21) and (1.22) to obtain the final form of the boundary conditions when there is discontinuity in the medium electrical or magnetic properties. Now, consider two media separated by interface S as shown in Fig. 1.3.

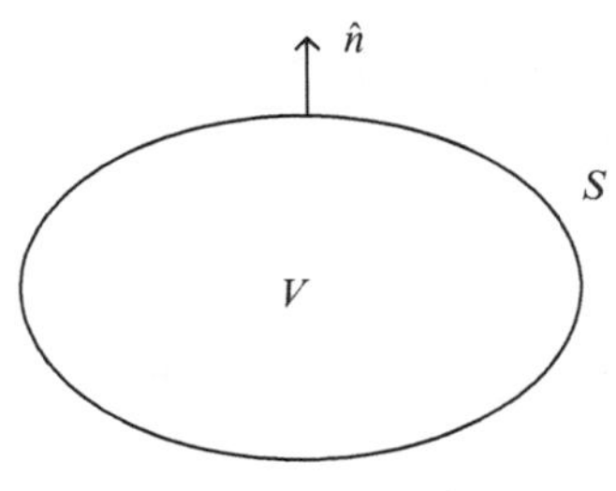

Fig. 1.2 Illustration of the divergence theorem

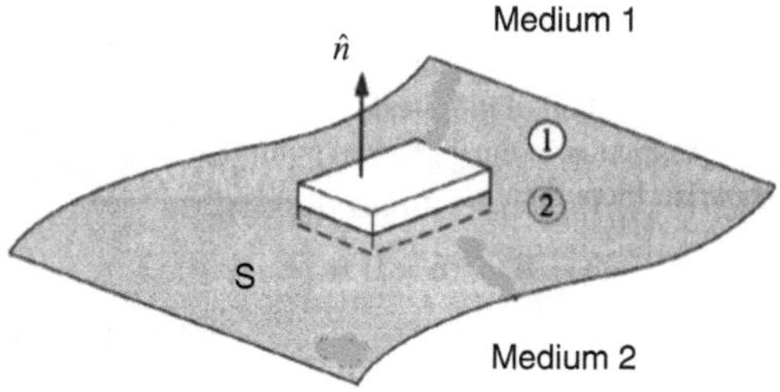

Fig. 1.3 Illustration of boundary conditions for normal components

If we apply (1.22) over the surface of the box shown above which is normal to the interface and assume the height of the box to be infinitely short, then the field may be considered constant over each face. It can then be shown that

$$(\bar{B}_1 - \bar{B}_2).\hat{n} = 0 \tag{1.23}$$

As a result, (1.23) indicates that the normal components of $\bar{B}$ is continuous across the boundary. If we apply same principle to (1.21), we obtain

$$(\bar{D}_1 - \bar{D}_2).\hat{n} = \rho_s \tag{1.24}$$

where ρ_s represents surface charge density. Equation (1.24) shows that the discontinuity in the normal components of $\bar{D}$ across the boundary is equal to the magnitude of surface charge density, ρ_s.

Now, consider thin rectangular loop across the boundary of the two medium shown in Fig. 1.4.

Applying (1.17) to sufficiently small rectangular loop shown above, then the field components can be considered constant. Moreover, ff the shorter sides of the loop are assumed to be infinitely small, then they don't contribute to the loop integral and as a result surface integral on the right hand side of the equation vanishes. The result can be expressed as

$$(\bar{E}_1 - \bar{E}_2).\hat{l} = 0 \tag{1.25}$$

Equation (1.25) can also be written using normal vector as

$$\hat{n} \times (\bar{E}_1 - \bar{E}_2) = 0 \tag{1.26}$$

Equations (1.25) or (1.26) simply states that the tangential components of $\bar{E}$ are continuous across the boundary. If we apply same principle to (1.18), we obtain

$$\hat{n} \times (\bar{H}_1 - \bar{H}_2) = \bar{J}_s \tag{1.27}$$

where $\bar{J}_s$ represents surface current density. Equation (1.24) shows that the discontinuity in the tangential components of $\bar{H}$ is equal to surface current density.

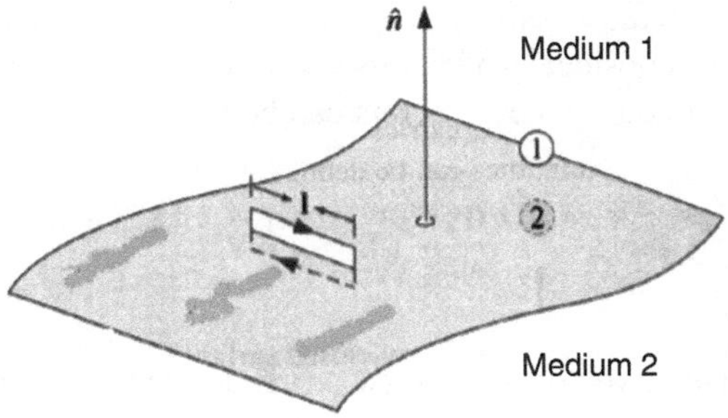

Fig. 1.4 Illustration of boundary conditions for tangential components

If the current density is finite as it is the case in any medium, then (1.27) can be written as

$$\hat{n} \times (\bar{H}_1 - \bar{H}_2) = 0 \tag{1.28}$$

Equation (1.28) states that tangential components of $\bar{H}$ across the boundary are continuous when the surface current density is finite.

1.4 Tensors and Dyadic Analysis

Vectors are essential in mathematical study of physical phenomena. Tensors can be considered as natural generalization of vectors. Consider (1.9) for isotropic medium. It is written as

$$\bar{D} = \varepsilon \bar{E} \tag{1.29}$$

Permittivity ε is a constant in this instance. As a result, electric flux density and electric field intensity might be different in the amplitude but they have the same direction. They are both vectors. However, when the medium is anisotropic and expressed as in (1.9),

$$\bar{D} = \varepsilon_0 \bar{\bar{\varepsilon}} \cdot \bar{E} \tag{1.30}$$

the direction of flux density and field intensity are no longer in the same direction. $\bar{D}$ and $\bar{E}$ are still vectors although their dimensions changed. The permittivity of the medium because is not constant anymore. It is in the form of a matrix and has nine components as shown in (1.11). At this point, we introduce the term *tensor* and call $\bar{\bar{\varepsilon}}$ as *permittivity tensor*. We use double over bar over the symbol to signify that the quantity is a tensor. In tensor notation, *scalars* are *tensor of rank 0, vectors* are *tensor of rank 1* and *matrices* are *tensors of rank 2*.

Dyad is simply product of a two vectors. Not dot product nor a cross product. If we consider two vectors $\bar{a} = a_1\hat{x} + a_2\hat{y} + a_3\hat{z}$ and $\bar{c} = c_1\hat{x} + c_2\hat{y} + c_3\hat{z}$, then a dyad formed by these vectors can be written as

$$\begin{aligned}\bar{a}\bar{c} = &a_1c_1\hat{x}\hat{x} + a_1c_2\hat{x}\hat{y} + a_1c_3\hat{x}\hat{z} + a_2c_1\hat{y}\hat{x} + a_2c_2\hat{y}\hat{y} + a_2c_3\hat{y}\hat{z} \\ &+a_3c_1\hat{z}\hat{x} + a_3c_2\hat{z}\hat{y} + a_3c_3\hat{z}\hat{z}\end{aligned} \tag{1.31}$$

Here, vector $\bar{a}$ is called antecedent and vector $\bar{c}$ is called consequent. The scalar components of all dyads can be written as a matrix as shown below.

$$\begin{bmatrix} a_1c_1 & a_1c_2 & a_1c_3 \\ a_2c_1 & a_2c_2 & a_2c_3 \\ a_3c_1 & a_3c_2 & a_3c_3 \end{bmatrix}$$

The sum of two or more dyads is called *dyadic*. It is important to note that all tensors of rank 2 are either dyad or matrix.

We need to review some of the critical properties about tensors that will be used in the following chapters:

- Dyad does not have commutative property, i.e.;

$$\bar{a}\bar{c} \neq \bar{c}\bar{a} \tag{1.32}$$

- The product of a dyad and a scalar is written as

$$\alpha(\bar{a}\bar{c}) = (\bar{a}\bar{c})\alpha = (\alpha\bar{a})\bar{c} = (\bar{a}\alpha)\bar{c} = \bar{a}(\alpha\bar{c}) = \bar{a}(\bar{c}\alpha) \tag{1.33}$$

- If $\bar{e}$ is a vector, then the inner product of a vector and dyad can be related as

$$\bar{e} \cdot (\bar{a}\bar{c}) = (\bar{e} \cdot \bar{a})\bar{c} = \lambda\bar{c} \tag{1.34}$$

where λ is a constant. The result of inner product of a vector and a dyad is a scaled vector.

- The trace of a dyad is basically dot product of vectors forming the dyad.

$$(\bar{a}\bar{c})_t = \bar{a} \cdot \bar{c} \tag{1.35}$$

- The determinant and adjoint of a dyad are zero.

$$|\bar{a}\bar{c}| = 0 \tag{1.36}$$

$$adj(\bar{a}\bar{c}) = 0 \tag{1.37}$$

The following example will help to clarify the concept of dyad even further. Consider (1.9) again. If we use the property given in (1.34), then we can express it as

$$\bar{D} = \varepsilon_o \bar{\bar{\varepsilon}} \cdot \bar{E} = \varepsilon_o (\bar{a}\bar{c}) \cdot \bar{E} = \varepsilon_o \bar{a}(\bar{c} \cdot \bar{E}) = \bar{a}\lambda \tag{1.38}$$

It is clear the direction of $\bar{D}$ is independent of the direction of $\bar{E}$. It only depends on the direction of vector $\bar{a}$.

The relation between matrices and dyads can be best understood by studying unit matrix. Any unit matrix can be decomposed into sum of three dyads using *completeness relation* as

$$\bar{\bar{I}} = \bar{a}_1\bar{a}_2 + \bar{b}_1\bar{b}_2 + \bar{c}_1\bar{c}_2 \tag{1.39}$$

where $\bar{a}_1$, $\bar{b}_1$, and $\bar{c}_1$ are three linearly independent, non-coplanar vectors and $\bar{a}_2$, $\bar{b}_2$, and $\bar{c}_2$ are three linearly independent vector set which are reciprocal to $\bar{a}_1$, $\bar{b}_1$, and $\bar{c}_1$. At this point we learned that we can express matrices in terms of dyads by following certain rules. Matrices can be categorized based on the dyadic form that they can be expressed. They can be *zero*, *complete*, *planar*, or *linear*:

- A matrix is considered to be a zero matrix if

$$\bar{A} = 0 \tag{1.40}$$

$$|\bar{A}| = 0 \tag{1.41}$$

$$adj(\bar{A}) = 0 \tag{1.42}$$

- A matrix that maps a three dimensional vector into another three dimensional vector is said to be complete. If $\bar{A}$ is complete matrix, then it has the following properties

$$\bar{A} = \bar{a}_1\bar{a}_2 + \bar{b}_1\bar{b}_2 + \bar{c}_1\bar{c}_2 \tag{1.43}$$

$$adj(\bar{A}) = \bar{k}_1\bar{k}_2 + \bar{l}_1\bar{l}_2 + \bar{m}_1\bar{m}_2 \tag{1.44}$$

$$|\bar{A}| \neq 0 \tag{1.45}$$

- A matrix is planar if

$$\bar{A} = \bar{a}_1\bar{a}_2 + \bar{b}_1\bar{b}_2 \tag{1.46}$$

$$adj(\bar{A}) = (\bar{a}_2 \times \bar{b}_2)(\bar{a}_1 \times \bar{b}_1) \tag{1.47}$$

$$|\bar{A}| = 0 \tag{1.48}$$

- A matrix is linear if

$$\bar{A} = \bar{a}_1\bar{a}_2 \tag{1.49}$$

$$adj(\bar{A}) = 0 \tag{1.50}$$

$$|\bar{A}| = 0 \tag{1.51}$$

1.5 Eigenvalue Problems

The understanding of the solution of the eigenvalue problems will greatly facilitate the solution of electromagnetic problems involving anisotropic or gyrotropic media. Assume, the matrix $\bar{A}$ is a square matrix of order n and $\bar{c}$ is an n dimensional vector. Then, the eigenvalue problem is given by

$$\bar{A}.\bar{c} = \alpha\bar{c} \tag{1.52}$$

where α is a constant. Equation (1.52) can be interpreted as follows. When a vector is multiplied with a matrix, it results in a scaling of the same vector by α. In this problem, vector $\bar{c}$ is *eigenvector* of matrix $\bar{A}$, and α is an *eigenvalue* of $\bar{A}$ for the corresponding eigenvector $\bar{c}$. The eigenvalue problem in (1.52) can be re-written as

$$(\bar{A} - \alpha\bar{I}).\bar{c} = 0 \tag{1.53}$$

$(\bar{A} - \alpha\bar{I})$ is called *characteristic matrix* of $\bar{A}$. Equation (1.53) is satisfied when vector $\bar{c} = 0$ for any value of eigenvalue α or when vector $\bar{c} \neq 0$ for the values of eigenvalue α which will satisfy (1.53). This is the solution we are looking which is nontrivial. This solution is given by the following equation.

$$|\bar{A} - \alpha\bar{I}| = 0 \tag{1.54}$$

$|\bar{A} - \alpha\bar{I}|$ is called characteristic equation of matrix $\bar{A}$. The eigenvalues of this characteristic equation is given by

$$D(\alpha) = |\bar{A} - \alpha\bar{I}| = \begin{vmatrix} a_{11} - \alpha & a_{12} & \dots & a_{1n} \\ a_{21} & a_{22} - \alpha & \dots & a_{2n} \\ \vdots & & & \vdots \\ a_{n1} - \alpha & a_{n2} & \dots & a_{nn} - \alpha \end{vmatrix} = 0 \tag{1.55}$$

$D(\alpha)$ is a polynomial of degree n in α. Equation (1.55) can be expanded as

$$D(\alpha) = v_0\alpha^n + v_1\alpha^{n-1} + v_2\alpha^{n-2} + \cdots + v_{n-1}\alpha + v_n \tag{1.56}$$

The coefficients v_j in (1.56) are functions of $\bar{A}$. We have to note that although every matrix of order n has n eigenvalues, it does necessarily have n linearly independent eigenvectors. This happens when some of the eigenvalues are repeated more than once. This is the case when (1.56) has two or more identical roots. If there are p repeated eigenvalues, then the matrix has at most k linearly independent eigenvectors all corresponding to same eigenvalue. These eigenvectors are known as *degenerate eigenvectors*. The rank of the matrix is then equal to $r = \text{n} - \text{p}$. As a result, the eigenvalue problem shown in (1.53) will have n − r linearly independent solutions. In our applications, we will deal with 3 × 3 matrices. When the dimension of $\bar{A}$ is 3 × 3, (1.55) can also be written in the form of

$$\alpha^3 - \bar{A}_t\alpha^2 + (adj\bar{A})_t\alpha - |\bar{A}| = 0 \tag{1.57}$$

There are four types of matrix $\bar{A}$ that are important in our analysis. They are when $\bar{A}$ is isotropic, uniaxial, biaxial, and gyrotropic. When $\bar{A}$ is symmetric, it can take isotropic, uniaxial or biaxial form. Real symmetric matrix also has *orthogonal* eigenvectors corresponding to distinct eigenvalues. When they are normalized by dividing them with their magnitude, we obtain *orthonormal* eigenvectors. When the matrix is Hermitan, it can take a gyrotropic form. Let's assume $\bar{A}$ is 3 × 3 real symmetric matrix. Then, Isotropic form of $\bar{A}$ is obtained when it has degeneracy of three, and rank of 0. This is possible when eigenvalues are $\alpha_1 = \alpha_2 = \alpha_3$. It can be shown that it can be written as

$$\bar{A} = \alpha\bar{I} \tag{1.58}$$

- Uniaxial form of $\bar{A}$ is obtained when it has degeneracy of two, and rank of 1. Its eigenvalues are $\alpha_1 = \alpha_2 \neq \alpha_3$. It can be shown that $\bar{A}$ can be expressed in dyadic form as

$$\bar{A} = \alpha_1(\hat{u}_1\hat{u}_1 + \hat{u}_2\hat{u}_2) + \alpha_3\hat{u}_3\hat{u}_3 \tag{1.59}$$

$\hat{u}_1, \hat{u}_2$ and $\hat{u}_3$ are the orthonormal eigenvectors. Equation (1.59) can be put in the following dyadic form.

$$\bar{A} = \alpha_1\bar{I} + \alpha_3\hat{u}_3\hat{u}_3 \tag{1.60}$$

where $\bar{I} = \hat{u}_1\hat{u}_1 + \hat{u}_2\hat{u}_2 + \hat{u}_3\hat{u}_3$. $\hat{u}_3$ refers the optic axis of the medium when $\bar{A}$ is the permittivity tensor of the anisotropic medium.

- Biaxial form of $\bar{A}$ is obtained when it has no degeneracy, and as a result has a rank of 3. Its eigenvalues are $\alpha_1 < \alpha_2 < \alpha_3$. It can be shown that $\bar{A}$ can be expressed in dyadic form as

$$\bar{A} = \alpha_2(\hat{u}_1\hat{u}_1 + \hat{u}_2\hat{u}_2 + \hat{u}_3\hat{u}_3) + (\alpha_3 - \alpha_2)\hat{u}_3\hat{u}_3 - (\alpha_2 - \alpha_1)\hat{u}_1\hat{u}_1 \tag{1.61}$$

Equation (1.61) can be further simplified and put in the following dyadic form.

$$\bar{A} = \alpha_2\bar{I} + \frac{1}{2}(\alpha_3 - \alpha_1)(\hat{s}\hat{r} + \hat{r}\hat{s}) \tag{1.62}$$

where $\hat{s}$ and $\hat{r}$ are unit vectors and related to the orthonormal eigenvectors with the following equations

$$\hat{s} = \left(\frac{\alpha_3 - \alpha_2}{\alpha_3 - \alpha_1}\right)\hat{u}_3 + \left(\frac{\alpha_2 - \alpha_1}{\alpha_3 - \alpha_1}\right)\hat{u}_1 \tag{1.63}$$

$$\hat{r} = \left(\frac{\alpha_3 - \alpha_2}{\alpha_3 - \alpha_1}\right)\hat{u}_3 - \left(\frac{\alpha_2 - \alpha_1}{\alpha_3 - \alpha_1}\right)\hat{u}_1 \tag{1.64}$$

- Gyrotropic form of $\bar{A}$ is obtained when it is Hermitian. General dyadic form of $\bar{A}$ can be given as

$$\bar{A} = \alpha_1\hat{u}_1\hat{u}_1^* + \alpha_2\hat{u}_2\hat{u}_2^* + \alpha_3\hat{u}_3\hat{u}_3^* \tag{1.65}$$

where α_1, α_2, and α_3 are the eigenvalues and $\hat{u}_1, \hat{u}_2$ and $\hat{u}_3$ are the orthonormal eigenvectors. The symbol * indicates the complex conjugate.

1.6 k-Domain Method

For monochromatic plane waves in homogenous media, plane wave solutions have the variation on space and time with $e^{i(\bar{k}\cdot\bar{r}-\omega t)}$ where $\bar{k}$ is the wave vector. k-domain analysis is obtained by assuming the plane wave solution is of the form

$$\bar{E}_{s,t}(\bar{r}, t) = \bar{E}e^{i(\bar{k}\cdot\bar{r}-\omega t)} \tag{1.66}$$

$\bar{E}$ is a constant vector. Then, we can introduce the following relation which will transform the problem under consideration into k-domain as

$$\nabla \rightarrow i\bar{k} \tag{1.67a}$$

$$\frac{\partial}{\partial t} \to -i\omega \tag{1.67b}$$

Maxwell equations in the source free region will take the following forms under the transformation shown in (1.67).

$$\bar{k} \times \bar{E} = \omega \bar{B} \tag{1.68}$$

$$\bar{k} \times \bar{H} = -\omega \bar{D} \tag{1.69}$$

$$\bar{k} \cdot \bar{B} = 0 \tag{1.70}$$

$$\bar{k} \cdot \bar{D} = 0 \tag{1.71}$$

For instance, the solution for vector $\bar{E}$ in isotropic medium using the equations given by (1.68)–(1.71) is easily obtained by cross multiplying (1.68) with vector $\bar{k}$ and use the constitutive relation given in (1.10) and the vector relation

$$\bar{a} \times (\bar{b} \times \bar{c}) = (\bar{a} \cdot \bar{c})\bar{b} - (\bar{a} \cdot \bar{b})\bar{c} \tag{1.72}$$

Then, we have

$$\bar{k} \times (\bar{k} \times \bar{E}) = \omega\mu\bar{k} \times \bar{H} \tag{1.73}$$

and substitute (1.69) into (1.73) with constitutive relation (1.9) and use the vector relation in (1.72) as

$$(\bar{k} \cdot \bar{E})\bar{k} - (\bar{k} \cdot \bar{k})\bar{E} = \omega\mu(-\omega\varepsilon\bar{E}) \tag{1.74}$$

Since $\bar{k} \cdot \bar{E} = 0$ from (1.71), then (1.74) can be simplified as

$$\left(k^2 - \omega^2\mu\varepsilon\right)\bar{E} = 0 \tag{1.75}$$

The non-trivial solution of (1.75) results in obtaining dispersion relation in isotropic medium as

$$k^2 = \omega^2\mu\varepsilon \tag{1.76}$$

One of the tools that we will be using in the k-domain is n-dimensional *Fourier Convolution Theorem*. This theorem will be used to relate spatial Green's functions and k-domain Green's functions. Assume, we are convoluting two functions $f(x)$ and $g(x)$. The convolution theorem for these two functions can be written as

$$\Im[f \circ g(x); x \to \xi] = F_{(n)}(\xi)G_{(n)}(\xi) \tag{1.77}$$

In (1.77), $F_{(n)}(\xi)$ and $G_{(n)}(\xi)$ are the Fourier transforms of $f(x)$ and $g(x)$. The inverse of (1.77) has also major application. It can be written as

$$\Im_{(n)}{}^{-1}\left[F_{(n)}(\xi)G_{(n)}(\xi); \xi \to x\right] = \frac{1}{(2\pi)^{0.5n}} \int_{E_n} f(x-u)g(u)du \tag{1.78}$$

Equation (1.78) will be used in the derivation of dyadic Green's functions in the k-domain in Chap. 4.

References

1. I.Y. Hsia, H.Y. Yang and N.G. Alexopoulos, "Basic properties of microstrip circuit elements on nonreciprocal substrate-superstrate structures," *J. Electromagn. Waves and Appl.*, vol. 5, no. 4/5, pp. 465–476, 1991.
2. I.Y. Hsia, H.Y. Yang and N.G. Alexopoulos, "Fundamental characteristics of microstrip antennas on ferrite semiconductor interfaces," in Proc. Int. Conf. Electromagn. In Aerosp. Appl., Torino, Italy, pp. 169–171, Sept. 12–15, 1989.
3. J. Helszajn, *Ferrite Phase Shifters and Control Devices*. San Francisco, CA: McGraw-Hill, 1989.
4. R.F. Soohoo, *Microwave Magnetics*. San Francisco, CA: Harper & Row, 1985, Chapter 9.
5. B. Lax and K.J. Button, *Microwave Ferrites and Ferrimagnetics*. New York: McGraw-Hill, 1962, Chapter 12.
6. J.D. Adam, L.E. Davis, G.F. Dionne, E.F. Schloemann, and S.N. Stitzer, "Ferrite devices and materials," *IEEE Trans. Microw. Theory Tech.*, vol. MTT-50, pp. 721 – 737, 2002.
7. B.R. McLeod and W.G. May, "A 35 GHz isolator using a coaxial solid state plasma in a longitudinal magnetic field," *IEEE Trans. Microw. Theory Tech.*, vol. MTT-24, no. 4, pp.201–208, 1976.
8. R. Sloan, C.K. Yong, and L.E. Davis, "Broadband theoretical gyroelectric junction circulator tracking behavior at 77 K," *IEEE Trans. Microw. Theory Tech.*, vol. 44, pp.2655–2660, 1996.
9. V.H. Mok and L.E. Davis, "Non-reciprocal GaAs phase shifters and isolators for millimetric and sub-millimetric wavelengths," *IEEE MTTS Int. Microw. Symp.*, vol. 3, 2003, pp. 2249–2252.

Chapter 2
Wave Propagation and Dispersion Characteristics in Anisotropic Medium

In this chapter, the general dispersion and constitutive relations for anisotropic medium are derived. The detailed analysis of wave propagation and the dispersion characteristics of uniaxially, biaxially anisotropic medium is given. Plane waves in anisotropic media are studied.

2.1 Dispersion Relations and Wave Matrices

The medium is called *anisotropic* when the electrical and/or magnetic properties of a medium depend upon the directions of field vectors. The relationships between fields can be written in the following form:

$$\bar{D} = \varepsilon_0 \bar{\bar{\varepsilon}} \cdot \bar{E} \tag{2.1}$$

$$\bar{B} = \mu_0 \bar{\bar{\mu}} \cdot \bar{H} \tag{2.2}$$

where $\bar{\bar{\varepsilon}}$ and $\bar{\bar{\mu}}$ are relative permittivity and permeability tensors, respectively.

Anisotropic materials may be divided into two classes, depending on whether the natural modes of propagation are linearly polarized or circularly polarized waves. In the former, the permittivity and permeability components are symmetric; that is $\varepsilon_{ik} = \varepsilon_{ki}$ and $\mu_{ik} = \mu_{ki}$. For the latter, called *gyrotropic* media, the permittivity or permeability components for lossless media are antisymmetric, having $\varepsilon_{ij} = -\varepsilon_{ji}$ or $\mu_{ij} = -\mu_{ji}$. In this chapter, we will be discussing the former class of anisotropic where there exists symmetry in the permittivity tensor.

A. Eroglu, *Wave Propagation and Radiation in Gyrotropic and Anisotropic Media*,
DOI 10.1007/978-1-4419-6024-5_2, © Springer Science+Business Media, LLC 2010

2.2 General Form of Dispersion Relations and Wave Matrices

Let's re-write Maxwell's equations in the k-domain as

$$\bar{k} \times \bar{E} = \omega \bar{B} \tag{2.3}$$

$$\bar{k} \times \bar{H} = -\omega \bar{D} \tag{2.4}$$

$$\bar{k} \cdot \bar{B} = 0 \tag{2.5}$$

$$\bar{k} \cdot \bar{D} = 0 \tag{2.6}$$

By observing (2.1)–(2.6), it is seen that wave vector $\bar{k}$ is perpendicular to both $\bar{B}$ and $\bar{D}$. In Chap. 1, we found the solution for a vector $\bar{E}$ in isotropic medium. In this case, we would like to find the solution for vector $\bar{E}$ in the anisotropic medium. At this point, we define an anti symmetric matrix, or a tensor, which will be used in our derivation as

$$\bar{k} \times \bar{I} = \bar{\bar{k}} = \begin{bmatrix} 0 & -k_z & k_y \\ k_z & 0 & -k_x \\ -k_y & k_x & 0 \end{bmatrix} \tag{2.7}$$

$\bar{\bar{k}}$ has the following property because it is anti symmetric matrix

$$\bar{\bar{k}}.\bar{P} = \bar{\bar{k}} \times \bar{P} \tag{2.8}$$

where $\bar{P}$ is an arbitrary vector. Then, repeating the same procedure in Chap. 1, we obtain the following equations in anisotropic medium

$$\bar{\bar{k}} \cdot \bar{E} = \omega \mu_0 (\bar{\bar{\mu}} \cdot \bar{H}) \tag{2.9}$$

$$\bar{\bar{k}} \cdot \bar{H} = -\omega \varepsilon_0 (\bar{\bar{\varepsilon}} \cdot \bar{E}) \tag{2.10}$$

From (2.9), we obtain

$$\frac{\bar{\bar{\mu}}^{-1} \cdot \bar{\bar{k}} \cdot \bar{E}}{\mu_0 \omega} = \bar{H} \tag{2.11}$$

Substitution of (2.11) into (2.10) gives

$$\frac{\bar{\bar{k}} \cdot \bar{\bar{\mu}}^{-1} \cdot \bar{\bar{k}} \cdot \bar{E}}{\mu_0 \omega} = -\omega \varepsilon_0 (\bar{\bar{\varepsilon}} \cdot \bar{E})$$

or

$$\bar{\bar{k}} \cdot \bar{\bar{\mu}}^{-1} \cdot \bar{\bar{k}} \cdot \bar{E} + k_0^2 \bar{\bar{\varepsilon}} \cdot \bar{E} = 0$$

which can be simplified to give

$$\left[\bar{\bar{k}} \cdot \bar{\bar{\mu}}^{-1} \cdot \bar{\bar{k}} + k_0^2 \bar{\bar{\varepsilon}} \right] \cdot \bar{E} = 0 \tag{2.12}$$

where $k_0^2 = \omega^2 \mu_0 \varepsilon_0$. We repeat the same technique to obtain the solution for vector $\bar{H}$. Using (2.10), we obtain

$$-\frac{\bar{\bar{\varepsilon}}^{-1} \cdot \bar{\bar{k}} \cdot \bar{H}}{\varepsilon_0 \omega} = \bar{E} \tag{2.13}$$

Substitution of (2.13) into (2.9) gives

$$-\frac{\bar{\bar{k}} \cdot \bar{\bar{\varepsilon}}^{-1} \cdot \bar{\bar{k}} \cdot \bar{H}}{\varepsilon_0 \omega} = \omega \mu_0 (\bar{\bar{\mu}} \cdot \bar{H})$$

or

$$\left[\bar{\bar{k}} \cdot \bar{\bar{\varepsilon}}^{-1} \cdot \bar{\bar{k}} + k_0^2 \bar{\bar{\mu}} \right] \cdot \bar{H} = 0 \tag{2.14}$$

We let

$$\bar{\bar{W}}_E = \left[\bar{\bar{k}} \cdot \bar{\bar{\mu}}^{-1} \cdot \bar{\bar{k}} + k_0^2 \bar{\bar{\varepsilon}} \right] \tag{2.15}$$

$$\bar{\bar{W}}_M = \left[\bar{\bar{k}} \cdot \bar{\bar{\varepsilon}}^{-1} \cdot \bar{\bar{k}} + k_0^2 \bar{\bar{\mu}} \right] \tag{2.16}$$

and re-write (2.12) and (2.14) as

$$\bar{\bar{W}}_E \cdot \bar{E} = 0 \tag{2.17}$$

$$\bar{\bar{W}}_M \cdot \bar{H} = 0 \tag{2.18}$$

In (2.17) and (2.18), $\bar{\bar{W}}_E$ and $\bar{\bar{W}}_M$ are called *electric wave matrix* and *magnetic wave matrix*, respectively. Non-trivial solutions of the field vectors, $\bar{E}$ and $\bar{H}$, are obtained only when

$$\left| \bar{\bar{W}}_E \right| = \left| \bar{\bar{k}} \cdot \bar{\bar{\mu}}^{-1} \cdot \bar{\bar{k}} + k_0^2 \bar{\bar{\varepsilon}} \right| = 0 \tag{2.19}$$

$$\left|\bar{\bar{W}}_H\right| = \left|\bar{\bar{k}} \cdot \bar{\bar{\varepsilon}}^{-1} \cdot \bar{\bar{k}} + k_0^2 \bar{\bar{\mu}}\right| = 0 \tag{2.20}$$

Equations (2.19) and (2.20) are the *dispersion relations* for general anisotropic medium. It is interesting to note the duality between the electric wave matrix and magnetic wave matrix. If one of the wave matrix is known, the other one can be simply obtained by using the following duality relations

$$\bar{\bar{\varepsilon}} \to \bar{\bar{\mu}}, \quad \mu \to \varepsilon, \quad \mu_0 \leftrightarrow \varepsilon_0. \tag{2.21}$$

2.2.1 Disperison Relation and Wave Matrix for Uniaxially Anisotropic Medium

The general wave matrices and dispersion relations can be used to obtain their corresponding forms for uniaxially anisotropic medium by substituting the following permittivity and permeability tensors into (2.15)

$$\bar{\bar{\varepsilon}} = \varepsilon_{11}\bar{\bar{I}} + (\varepsilon_{33} - \varepsilon_{11})\hat{p}\hat{p} \tag{2.22}$$

$$\bar{\bar{\mu}} = \bar{\bar{I}} \tag{2.23}$$

Please note that we are using double over bar to represent dyadics. $\hat{p}$ shows the direction of the optic axis exist in the uniaxially anisotropic medium and equals $\hat{p} = \hat{z}$. In matrix form, the permittivity tensor can be written as

$$\bar{\bar{\varepsilon}} = \begin{bmatrix} \varepsilon_{11} & 0 & 0 \\ 0 & \varepsilon_{11} & 0 \\ 0 & 0 & \varepsilon_{33} \end{bmatrix} \tag{2.24}$$

The wave numbers for uniaxially anisotropic medium is obtained by solving (2.17). We are looking for non-trivial solution that will satisfy (2.17). That is the case when $\left|\bar{\bar{W}}_E\right| = 0$. Substituting (2.22)–(2.23) into general electric wave matrix given in (2.15) gives the electric wave matrix for uniaxially anisotropic medium as

$$\bar{\bar{W}}_E = (k_0^2\varepsilon_{11} - k^2)\bar{\bar{I}} + \bar{k}\bar{k} + k_0^2(\varepsilon_{33} - \varepsilon_{11})\hat{z}\hat{z} \tag{2.25}$$

The adjoint matrix, $adj\left(\bar{\bar{W}}_E\right)$ in dyadic form can be obtained using the flowing identities

$$\bar{A} = \lambda\bar{\bar{I}} + \bar{u}\bar{v} + \bar{m}\bar{n} \tag{2.26}$$

$$|\bar{A}| = \lambda\left[\lambda^2 + \lambda(\bar{u}\cdot\bar{v} + \bar{m}\cdot\bar{n}) + (\bar{v}\times\bar{m})\cdot(\bar{u}\times\bar{m})\right] \tag{2.27}$$

$$adj(\bar{A}) = \lambda\left[(\lambda + \bar{u}\cdot\bar{v} + \bar{m}\cdot\bar{n})\bar{\bar{I}} - \bar{u}\bar{v} - \bar{m}\cdot\bar{n}\right] + (\bar{v}\times\bar{n})(\bar{u}\times\bar{m}) \tag{2.28}$$

So, the dispersion relation for unaxially anisotropic medium can be obtained as

$$\left|\bar{\bar{W}}_E\right| = k_0^2\left(k^2 - k_0^2\varepsilon_{11}\right)\left[(\bar{k}\cdot\bar{\bar{\varepsilon}}\cdot\bar{k}) - k_0^2\varepsilon_{11}\varepsilon_{33}\right] = 0 \tag{2.29}$$

Since $\bar{k} = \hat{k}k$, then the wave numbers are obtained from (2.29) as

$$k_I^2 = k_0^2\varepsilon_{11} \tag{2.30}$$

$$k_{II}^2 = \frac{k_0^2\varepsilon_{11}\varepsilon_{33}}{\hat{k}\cdot\bar{\bar{\varepsilon}}\cdot\hat{k}} \tag{2.31}$$

Wave numbers k_I^2 and k_{II}^2 represent two types of wave numbers propagating in uniaxially anisotropic medium. k_I represents ordinary waves whereas k_{II} represents extraordinary waves. k_I does not depend on the direction of the wave normal whereas k_{II} does. This dependency is more clear when (2.24) is substituted into (2.31). We obtain

$$k_{II}^2 = \frac{k_0^2\varepsilon_{11}\varepsilon_{33}}{\varepsilon_{33}\cos^2\theta + \varepsilon_{11}\sin^2\theta} \tag{2.32}$$

where θ represents the angle between the optic axis and wave vector $\bar{k}$ as shown below in Figs. 2.1 and 2.2.

The extraordinary wave turns into an ordinary wave when the medium is isotropic and the equation given in (2.30)–(2.32) become identical because the direction dependency of the wave normal for extraordinary wave is removed. In that case, $\varepsilon_{11} = \varepsilon_{22} = \varepsilon_{33}$ and (2.32) reduces to (2.30).

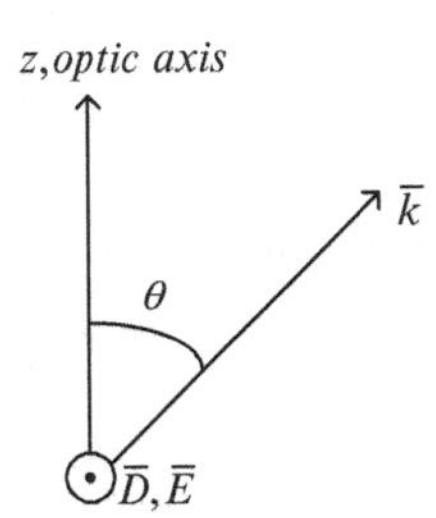

Fig. 2.1 Direction of wave vector and field vectors with respect to optic axis for an ordinary wave

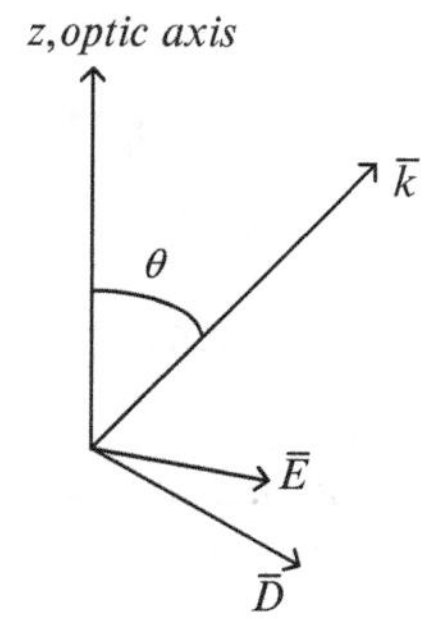

Fig. 2.2 Direction of wave vector and field vectors with respect to optic axis for an extraordinary wave

Equation (2.30) is an equation of a sphere with a radius $k_0\sqrt{\varepsilon_{11}}$. This is clearer when this equation is expanded as

$$k_x^2 + k_y^2 + k_z^2 = k_0^2\varepsilon_{11} \tag{2.33}$$

In analytical geometry, the equation for sphere is given as

$$(x - x_0)^2 + (y - y_0)^2 + (z - z_0)^2 = r^2 \tag{2.34}$$

where x_0, y_0, and z_0 define center point of the sphere. The transformation to spherical coordinate system is accomplished by the following equations

$$x = x_0 + r\sin\theta\cos\phi \tag{2.35a}$$

$$y = y_0 + r\sin\theta\sin\phi \tag{2.35b}$$

$$z = z_0 + r\cos\theta \tag{2.35c}$$

Where $0 < \theta < \pi$ and $0 < \phi < 2\pi$. Hence, the wave normal surface for ordinary waves in a uniaxial medium is defined by a sphere just like the waves in isotropic medium. Equation (2.31) is an equation for an ellipsoid. Equation (2.31) can be rewritten as

$$\frac{k_x^2}{\omega^2\mu\varepsilon_{33}} + \frac{k_y^2}{\omega^2\mu\varepsilon_{33}} + \frac{k_z^2}{\omega^2\mu\varepsilon_{11}} = 1 \tag{2.36}$$

Ellipsoid in analytical geometry is defined as

$$\frac{x^2}{a^2} + \frac{y^2}{b^2} + \frac{z^2}{c^2} = 1 \tag{2.37}$$

where a and b are the radii along the x and y axes and c is the polar radius along the z-axis. The more general form is obtained with the following relations.

$$x^T A x = 1 \tag{2.38}$$

where A is a symmetric positive definite matrix and $\mathbf{x}$ is a vector. In that case, the eigenvectors of A define the principal directions of the ellipsoid and the inverse of the square root of the eigenvalues are the corresponding equatorial radii. If (2.31) is carefully investigated, it can be put the following form.

$$\bar{k} \cdot \bar{\bar{\varepsilon}} \cdot \bar{k} = k_0^2 \varepsilon_{11} \varepsilon_{33}$$

or

$$\bar{k} \cdot \bar{\bar{\varepsilon}}_c \cdot \bar{k} = 1 \tag{2.39}$$

where

$$\bar{\bar{\varepsilon}}_c = \begin{bmatrix} \frac{1}{k_0^2 \varepsilon_{33}} & 0 & 0 \\ 0 & \frac{1}{k_0^2 \varepsilon_{33}} & 0 \\ 0 & 0 & \frac{1}{k_0^2 \varepsilon_{11}} \end{bmatrix} \tag{2.40}$$

Equation (2.40) is a symmetric positive definite matrix. The length of the principle axis along the transverse direction is $k_0\sqrt{\varepsilon_{33}}$ and $k_0\sqrt{\varepsilon_{11}}$ along k_z. Equations (2.30) and (2.31) can be used to obtain the wave surfaces for ordinary and extraordinary waves for negatively and positively uniaxially anisotropic media. The wave surface for negatively uniaxial medium is illustrated in Fig. 2.3 whereas the wave surface for positively uniaxial medium is illustrated in Fig. 2.4.

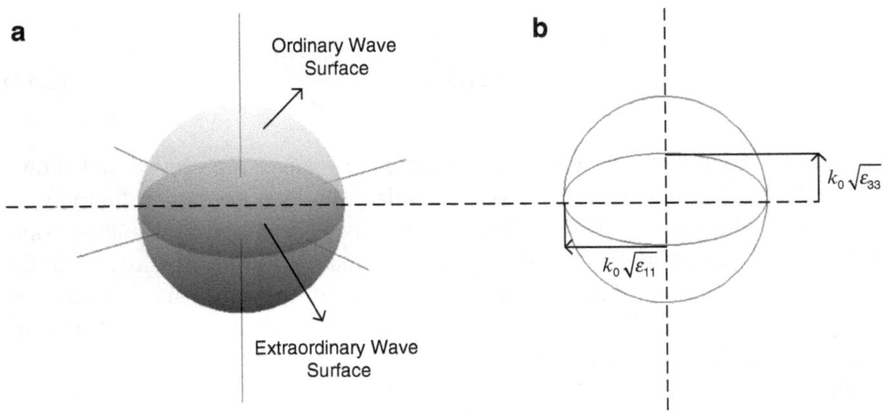

Fig. 2.3 Wave surface for negatively unaxial medium, $\varepsilon_{33} < \varepsilon_{11}$. (**a**) 3D plot ordinary wave and extraordinary wave surfaces (**b**) 2D plot of wave normal surfaces

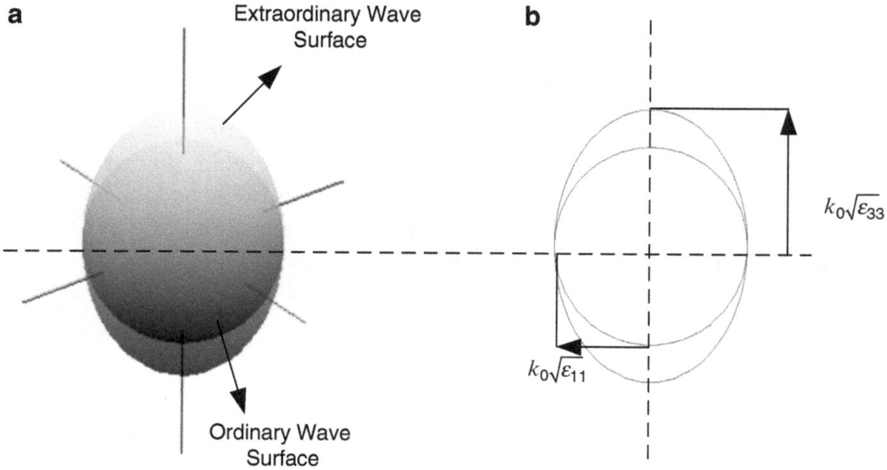

Fig. 2.4 Wave surface for positively unaxial medium, $\varepsilon_{33} > \varepsilon_{11}$. (**a**) 3D plot ordinary wave and extraordinary wave surfaces (**b**) 2D plot of wave normal surfaces

The phase velocity of the wave is defined by

$$v_p = \frac{\omega}{k} \tag{2.41}$$

The phase velocity of the ordinary wave is

$$v_{po} = \frac{\omega}{k_I} = \pm\frac{1}{\sqrt{\mu\varepsilon_{11}}} \tag{2.42}$$

where as the phase velocity for the extraordinary wave is equal to

$$v_{pe} = \frac{\omega}{k_{II}} = \pm\sqrt{\frac{\cos^2(\theta)}{\mu\varepsilon_{11}} + \frac{\sin^2(\theta)}{\mu\varepsilon_{33}}} \tag{2.43}$$

The phase velocities for ordinary and extraordinary waves are plotted and shown below. As it is illustrated, ordinary wave travels faster than the extraordinary wave when the medium is positively uniaxial. If it is negatively uniaxially anisotropic medium, this characteristic reverses and extraordinary wave travels faster than the ordinary wave. At the instances, when the direction of wave normal coincides with the optic axis, i.e., $\theta = 0°$, $180°$, etc., two waves degenerate into one and propagate at the same velocity as shown in Fig. 2.5.

The dispersion relation for uniaxially anisotropic medium given in (2.29) was expressed in terms of wave normals k_I and k_{II}. In the problems with the multilayer structures where the stratification of the layers is perpendicular, it is practical to use the wave number along that direction to simplify the analysis. For instance, if the

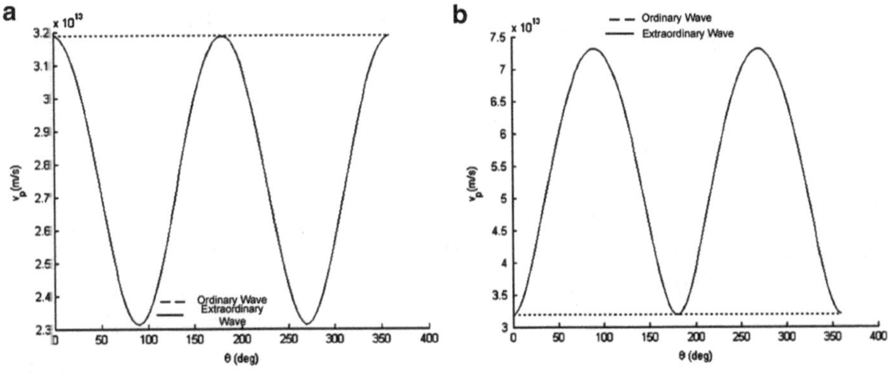

Fig. 2.5 Phase velocity response of ordinary and extraordinary waves for (**a**) positively uniaxial medium (**b**) negatively uniaxial medium

stratification of the layers is in z-direction then it is practical to use k_z as the wave number variable. The dispersion relation and k_z are obtained by using the following relations

$$k^2 = k_x^2 + k_y^2 + k_z^2 \tag{2.44}$$

$$k_\rho^2 = k_x^2 + k_y^2 = k^2 \sin^2\theta \tag{2.45}$$

$$k_z^2 = k^2 \cos^2\theta \tag{2.46}$$

When (2.44)–(2.46) substituted into (2.29), we obtain the dispersion relation for uniaxial anisotropic medium in terms of k_z as

$$\left|\bar{\bar{W}}_E\right| = k_0^2\left(k_\rho^2 + k_z^2 - k_0^2\varepsilon_{11}\right)\left[k_\rho^2\varepsilon_{11} + k_z^2\varepsilon_{33} - k_0^2\varepsilon_{11}\varepsilon_{33}\right] = 0 \tag{2.47}$$

Equation (2.47) has two positive roots in k_z^2. They are

$$k_{zI}^2 = k_0^2\varepsilon_{11} - k_\rho^2 \tag{2.48}$$

$$k_{zII}^2 = k_0^2\varepsilon_{11} - k_\rho^2\frac{\varepsilon_{11}}{\varepsilon_{33}} \tag{2.49}$$

where

$$k^2 = k_\rho^2 + k_z^2 = k_x^2 + k_y^2 + k_z^2$$

The adjoint of (2.25) which will be used in the dyadic Green's function derivation in the following chapters is obtained using the identity given in (2.28) as

$$adj\left(\bar{\bar{W}}_E\right) = \left(k_0^2\varepsilon_{11} - k^2\right)\left[k_0^2\varepsilon_{11}\bar{\bar{I}} - \bar{k}\bar{k} - k_0^2(\varepsilon_{33} - \varepsilon_{11})\hat{z}\hat{z}\right] + k_0^2(\varepsilon_{33} - \varepsilon_{11}) \times (\bar{k} \times \hat{z})(\bar{k} \times \hat{z}) \tag{2.50}$$

2.2.2 Disperison Relation and Wave Matrix for Biaxially Anisotropic Medium

Biaxial medium is defined by the following permittivity and permeability tensors.

$$\bar{\bar{\varepsilon}} = \varepsilon_{22}\bar{\bar{I}} + (\varepsilon_{33} - \varepsilon_{22})\hat{e}_3\hat{e}_3 - (\varepsilon_{22} - \varepsilon_{11})\hat{e}_1\hat{e}_1 \tag{2.51}$$

$$\bar{\bar{\mu}} = \bar{\bar{I}} \tag{2.52}$$

where $\hat{e}_1, \hat{e}_2$, and $\hat{e}_3$ are the unit vector and defined as

$$\hat{e}_1 = \begin{bmatrix} 1 \\ 0 \\ 0 \end{bmatrix}, \quad \hat{e}_2 = \begin{bmatrix} 0 \\ 1 \\ 0 \end{bmatrix}, \quad \hat{e}_3 = \begin{bmatrix} 0 \\ 0 \\ 1 \end{bmatrix} \tag{2.53}$$

The permittivity tensor in matrix form can be written as

$$\bar{\bar{\varepsilon}} = \begin{bmatrix} \varepsilon_{11} & 0 & 0 \\ 0 & \varepsilon_{22} & 0 \\ 0 & 0 & \varepsilon_{33} \end{bmatrix} \tag{2.54}$$

where $\varepsilon_{11} < \varepsilon_{22} < \varepsilon_{33}$. Equation (2.51) can be put in the following form.

$$\bar{\bar{\varepsilon}} = \varepsilon_{22}\bar{\bar{I}} + \left(\frac{\varepsilon_{33} - \varepsilon_{11}}{2}\right) \left\{ \begin{array}{l} \left(\left(\sqrt{\frac{\varepsilon_{33} - \varepsilon_{22}}{\varepsilon_{33} - \varepsilon_{11}}}\right)\hat{e}_3 + \left(\sqrt{\frac{\varepsilon_{22} - \varepsilon_{11}}{\varepsilon_{33} - \varepsilon_{11}}}\right)\hat{e}_1\right)\left(\left(\sqrt{\frac{\varepsilon_{33} - \varepsilon_{22}}{\varepsilon_{33} - \varepsilon_{11}}}\right)\hat{e}_3 - \left(\sqrt{\frac{\varepsilon_{22} - \varepsilon_{11}}{\varepsilon_{33} - \varepsilon_{11}}}\right)\hat{e}_1\right) \\ + \left(\left(\sqrt{\frac{\varepsilon_{33} - \varepsilon_{22}}{\varepsilon_{33} - \varepsilon_{11}}}\right)\hat{e}_3 - \left(\sqrt{\frac{\varepsilon_{22} - \varepsilon_{11}}{\varepsilon_{33} - \varepsilon_{11}}}\right)\hat{e}_1\right)\left(\left(\sqrt{\frac{\varepsilon_{33} - \varepsilon_{22}}{\varepsilon_{33} - \varepsilon_{11}}}\right)\hat{e}_3 + \left(\sqrt{\frac{\varepsilon_{22} - \varepsilon_{11}}{\varepsilon_{33} - \varepsilon_{11}}}\right)\hat{e}_1\right) \end{array} \right\} \tag{2.55}$$

In (2.55), let

$$\rho_1 = \sqrt{\frac{\varepsilon_{33} - \varepsilon_{22}}{\varepsilon_{33} - \varepsilon_{11}}}, \quad \rho_2 = \sqrt{\frac{\varepsilon_{22} - \varepsilon_{11}}{\varepsilon_{33} - \varepsilon_{11}}}, \quad \kappa_1 = \varepsilon_{22}, \quad \kappa_2 = \frac{\varepsilon_{33} - \varepsilon_{11}}{2} \tag{2.56}$$

and

$$\hat{h} = \rho_1 \hat{e}_3 + \rho_2 \hat{e}_1 \tag{2.57}$$

$$\hat{g} = \rho_1 \hat{e}_3 - \rho_2 \hat{e}_1 \tag{2.58}$$

Hence, (2.55) can be written as

$$\bar{\bar{\varepsilon}} = \kappa_1 \bar{\bar{I}} + \kappa_2 \left[\hat{h}\hat{g} + \hat{g}\hat{h} \right] \tag{2.59}$$

We substitute (2.59) into electric wave matrix given by (2.15) and obtain

$$\bar{\bar{W}}_E = \bar{k}\bar{k} - k^2 \bar{\bar{I}} + k_0^2 \left(\kappa_1 \bar{\bar{I}} + \kappa_2 \left[\hat{h}\hat{g} + \hat{g}\hat{h} \right] \right) \tag{2.60}$$

(2.60) can further be simplified as

$$\bar{\bar{W}}_E = \bar{k}\bar{k} + \left(k_0^2 \kappa_1 - k^2 \right) \bar{\bar{I}} + k_0^2 \kappa_2 \left[\hat{h}\hat{g} + \hat{g}\hat{h} \right] \tag{2.61}$$

Let

$$\begin{aligned} \bar{A} &= \left(k_0^2 \kappa_1 - k^2 \right) \bar{\bar{I}} + k_0^2 \kappa_2 \left[\hat{h}\hat{g} + \hat{g}\hat{h} \right] \\ &= \chi_1 \bar{\bar{I}} + \chi_2 \left[\hat{h}\hat{g} + \hat{g}\hat{h} \right] \end{aligned} \tag{2.62}$$

where

$$\chi_1 = \left(k_0^2 \kappa_1 - k^2 \right) \tag{2.63}$$

$$\chi_2 = k_0^2 \kappa_2 \tag{2.64}$$

Then, (2.61) can be re-written as

$$\bar{\bar{W}}_E = \bar{k}\bar{k} + \bar{A} \tag{2.65}$$

We need to use following identities for $\bar{\bar{W}}_E = \bar{A} + \bar{m}\bar{n}$ to obtain the dispersion relation.

$$\left| \bar{\bar{W}}_E \right| = |\bar{A}| + \bar{k} \cdot adj(\bar{A}) \cdot \bar{k} \tag{2.66}$$

$$adj\left(\bar{\bar{W}}_E \right) = adj(\bar{A}) + (\bar{A} - \bar{A}_t \bar{I}) \cdot (\bar{k} \times \bar{I}) \cdot (\bar{k} \times \bar{I}) + [(\bar{k} \cdot \bar{A}) \times \bar{I}] \cdot (\bar{k} \times \bar{I}) \tag{2.67}$$

In (2.62),

$$|\bar{A}| = \chi_1\left[(\chi_1 + \chi_2\hat{h}.\hat{g})^2 - \chi_2^2\right] \tag{2.68a}$$

$$\bar{A}_t = 3\chi_1 + 2\chi_2(\hat{h}.\hat{g}) \tag{2.68b}$$

$$adj(\bar{A}) = \chi_1\left[(\chi_1 + 2\chi_2\hat{h}.\hat{g})\bar{\bar{I}} - \chi_2(\hat{h}\hat{g} + \hat{g}\hat{h})\right] - \chi_2^2(\hat{h}\times\hat{g})(\hat{h}\times\hat{g}) \tag{2.68c}$$

Substituting (2.68) into (2.66) gives the final form of the dispersion relation for biaxial medium as

$$\begin{aligned} \left|\bar{\bar{W}}_E\right| &= \chi_1\left[(\chi_1 + \chi_2\hat{h}.\hat{g})^2 - \chi_2^2\right] \\ &+ \bar{k}\cdot\left[\chi_1\left[(\chi_1 + 2\chi_2\hat{h}.\hat{g})\bar{\bar{I}} - \chi_2(\hat{h}\hat{g} + \hat{g}\hat{h})\right] - \chi_2^2(\hat{h}\times\hat{g})(\hat{h}\times\hat{g})\right]\cdot\bar{k} \end{aligned} \tag{2.69}$$

2.3 Plane Waves in Anisotropic Medium

In the analysis of the plane wave propagation in anisotropic medium, we are looking for the monochromatic uniform plane wave solution with space and time dependence $e^{i(\bar{k}\cdot\bar{r}-\omega t)}$. Maxwell's equations expressed in (2.3) and (2.4) are re-written below show the relations between vectors $\bar{k}, \bar{D}$ and $\bar{H}$.

$$\bar{k}\times\bar{E} = \omega\bar{B} \tag{2.70}$$

$$\bar{k}\times\bar{H} = -\omega\bar{D} \tag{2.71}$$

It is seen from (2.70), (2.71) that the three vectors, $\bar{k}, \bar{D}$ and $\bar{H}$ are mutually perpendicular to each other. Vectors $\bar{D}, \bar{E}$ and $\bar{k}$ are co-planar and perpendicular to $\bar{H}$ because $\bar{H}$ is perpendicular to $\bar{E}$. The direction of power flow is perpendicular to both $\bar{E}$ and $\bar{H}$ and found from

$$\bar{S} = \bar{E}\times\bar{H} \tag{2.72}$$

The Poynting's vector in isotropic medium is in the direction of wave vector $\bar{k}$. This is not the case when the medium is anisotropic. However, $\bar{S}$ is co-planar with $\bar{D}, \bar{E}$ and $\bar{k}$. In addition it is important to note that the angle between $\bar{D}, \bar{E}$ and $\bar{k}, \bar{S}$ are equal. The relations between the field vectors, Poynthing's vector and wave vector can be illustrated in Fig. 2.6.

At this point, We introduce a refractive index vector $\bar{n}$ to physically better understand the analysis. It is defined as

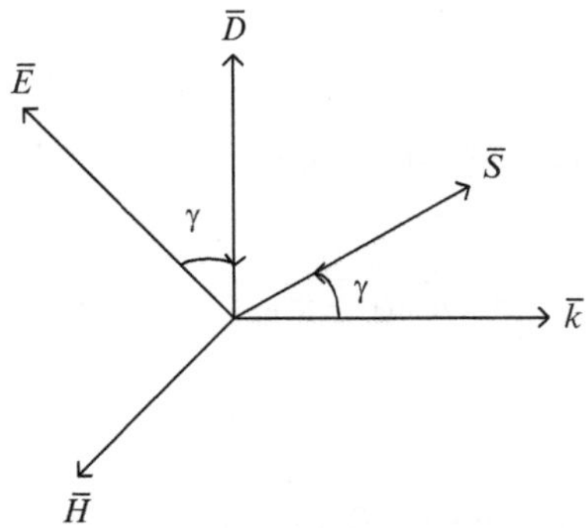

Fig. 2.6 The relations between vectors in aniostropic medium

$$\bar{n} = \frac{\bar{k}}{k_0} \tag{2.73}$$

Substituting (2.73) into (2.70) and (2.71) gives

$$\bar{n} \times \bar{E} = \sqrt{\frac{\mu_0}{\varepsilon_0}} \bar{\bar{\mu}} \cdot \bar{H} \tag{2.74}$$

$$\bar{n} \times \bar{H} = -\sqrt{\frac{\varepsilon_0}{\mu_0}} \bar{\bar{\varepsilon}} \cdot \bar{E} \tag{2.75}$$

Eliminating vector $\bar{H}$ from (2.74)–(2.75) and solving for vector $\bar{E}$ gives,

$$\bar{n} \times (\bar{n} \times \bar{E}) = -\bar{\bar{\varepsilon}} \cdot \bar{E} \tag{2.76}$$

Using the vector identity

$$\bar{a} \times (\bar{b} \times \bar{c}) = (\bar{a} \cdot \bar{c})\bar{b} - (\bar{a} \cdot \bar{b})\bar{c} \tag{2.77}$$

we can re-write (2.76) as

$$n^2\bar{E} - (\bar{n}.\bar{E})\bar{n} = \bar{D} = \bar{\bar{\varepsilon}} \cdot \bar{E} \tag{2.78}$$

Rearranging (2.78) gives the electric wave matrix in terms of refractive index vector as

$$\left[n^2\bar{\bar{I}} - \bar{n}\bar{n} - \bar{\bar{\varepsilon}}\right] \cdot \bar{E} = 0 \tag{2.79}$$

For non-zero solution of electric field vector, the determinant of the wave matrix above should be zero. It gives another general form of dispersion relation in terms of refractive index vector. It is shown as

$$\left| n^2 \bar{\bar{I}} - \bar{n}\bar{n} - \bar{\bar{\varepsilon}} \right| = 0 \tag{2.80}$$

When (2.80) is expanded,

$$n^2 \left(n_x^2 \varepsilon_{11} + n_y^2 \varepsilon_{22} + n_z^2 \varepsilon_{33} \right) - \begin{bmatrix} n_x^2 \varepsilon_{11}(\varepsilon_{22} + \varepsilon_{33}) + n_y^2 \varepsilon_{22}(\varepsilon_{11} + \varepsilon_{33}) \\ + n_z^2 \varepsilon_{33}(\varepsilon_{11} + \varepsilon_{22}) \end{bmatrix} + \varepsilon_{11}\varepsilon_{22}\varepsilon_{33} = 0 \tag{2.81}$$

Equation in (2.81) gives the wave vector surface for given permittivity tensor components and defines the magnitude of the wave vector. The solution gives two real values in n^2 for given direction of $\bar{n}$. The group velocity in the medium is given as

$$v_g = \frac{\partial \omega}{\partial k} \tag{2.82}$$

The direction of the group velocity vector defines the direction of the power flow in the medium.

Chapter 3
Wave Propagation and Dispersion Characteristics in Gyrotropic Medium

3.1 Introduction

In this chapter, we derive the general dispersion and constitutive relations for a gyrotropic medium. A detailed analysis of wave propagation in an electrically gyrotropic or gyroelectric medium will be given. We then obtain the dispersion relations in terms of angle θ, which is the angle between the wave normal and the external magnetic field, and in terms of the transverse component of the wave vector, k_ρ. We will analyze the plane waves in a gyroelectric medium and consider the cut off and resonance conditions for the *principle waves*. We then use the results to construct the *Clemmow-Mually-Allis* (*CMA*) diagram and tabulate the frequency bands over which the wave can propagate in each region on this diagram.

3.2 Constitutive Relations

The medium is called *anisotropic* when the electrical and/or magnetic properties of a medium depend upon the directions of field vectors. The relationships between fields can be written in the following form.

$$\begin{aligned} \bar{D} &= \varepsilon_0 \bar{\bar{\varepsilon}} \bullet \bar{E} \\ \bar{B} &= \mu_0 \bar{\bar{\mu}} \bullet \bar{H} \end{aligned} \tag{3.1}$$

where $\bar{\bar{\varepsilon}}$ and $\bar{\bar{\mu}}$ are relative permittivity and permeability tensors, respectively. ε_0 and μ_0 are defined as the permittivity and permeability of free space.

Anisotropic materials may be divided into two classes, depending on whether the natural modes of propagation are linearly polarized or circularly polarized waves. In the former, the permittivity and permeability components are symmetric; that is $\varepsilon_{ik} = \varepsilon_{ki}$ and $\mu_{ik} = \mu_{ki}$. For the latter, called *gyrotropic* media, the permittivity or permeability components for lossless media are antisymmetric,

A. Eroglu, *Wave Propagation and Radiation in Gyrotropic and Anisotropic Media*,
DOI 10.1007/978-1-4419-6024-5_3, © Springer Science+Business Media, LLC 2010

having $\varepsilon_{ij} = -\varepsilon_{ji}$ or $\mu_{ij} = -\mu_{ji}$. Gyrotropic behavior results from the application of finite magnetic field to a plasma, to a ferrite, and to some dielectric crystals. In the presence of constant magnetic field $\bar{B}_0$, the tensor $\bar{\bar{\varepsilon}}$ or in index notation ε_{ik} is no longer symmetrical. The generalized symmetry of the kinetic coefficients requires that

$$\varepsilon_{ik}(\bar{B}_0) = \varepsilon_{ki}(-\bar{B}_0) \tag{3.2}$$

The condition requires that there is no dissipation and the tensor should be Hermitian:

$$\varepsilon_{ik} = \varepsilon_{ki}{}^{*} \tag{3.3}$$

Equation (3.3) implies only that the real and imaginary parts of ε_{ik} must be respectively symmetrical and antisymmetrical:

$$\varepsilon_{ik}{}' = \varepsilon_{ki}{}', \qquad \varepsilon_{ik}{}'' = -\varepsilon_{ki}{}'' \tag{3.4}$$

Using (3.2), we have

$$\begin{aligned} \varepsilon'_{ik}(\overline{B}_0) &= \varepsilon'_{ki}(\overline{B}_0) = \varepsilon'_{ik}(-\overline{B}_0) \\ \varepsilon''_{ik}(\overline{B}_0) &= -\varepsilon''_{ki}(\overline{B}_0) = -\varepsilon''_{ik}(-\overline{B}_0) \end{aligned} \tag{3.5}$$

In a dissipationless or lossless medium ε'_{ik} is an even function of $\bar{B}_0$, and ε''_{ik} is an odd function. The inverse tensor $\varepsilon_{ik}{}^{-1}$ evidently has the same symmetry properties, and is more convenient for use in the following calculations. To simplify the notation we shall write

$$\varepsilon_{ik}^{-1} \equiv \eta_{ik} = \eta_{ik}{}' + i\eta_{ik}{}'' \tag{3.6}$$

Any antisymmetrical tensor of rank two is equivalent (dual) to some axial vector; let the vector corresponding to the tensor η''_{ik} be $\bar{G}$. Using the antisymmetrical unit tensor e_{ikl}, we can write the relation between the components η''_{ik} and G_i as

$$\eta''_{ik} = e_{ikl}G_l \tag{3.7}$$

or, in components, $\eta_{xy}{}'' = G_z, \eta_{zx}{}'' = G_y, \eta_{yz}{}'' = G_x$. The relation $E_i = \eta_{ik}D_k$ between the electrical field and the electric displacement becomes

$$E_i = (\eta'_{ik} + ie_{ikl}G_l)D_k = \eta'_{ik}D_k + i(\overline{D} \times \overline{G})_i \tag{3.8}$$

A medium in which the relation between $\bar{E}$ and $\bar{D}$ is of this form is said to be *gyrotropic*. Gyrotropic medium becomes *electrically gyrotropic* or *gyroelectric* if the medium is characterized by the relative permittivity tensor in the following dyadic form:

$$\bar{\bar{\varepsilon}} = \varepsilon_1(\bar{\bar{I}} - \hat{b}_0\hat{b}_0) + i\varepsilon_2(\hat{b}_0 \times \bar{\bar{I}}) + \varepsilon_3\hat{b}_0\hat{b}_0, \quad \bar{\bar{\mu}} = \mu_0\bar{\bar{I}} \tag{3.9}$$

where $\hat{b}_0$ shows the direction of the applied magnetic field $\bar{B}_0$. Gyrotropic medium becomes *magnetically gyrotropic* or *gyromagnetic* if the medium is characterized by the relative permeability tensor in the following dyadic form:

$$\bar{\bar{\mu}} = \mu_1(\bar{\bar{I}} - \hat{b}_0\hat{b}_0) + i\mu_2(\hat{b}_0 \times \bar{\bar{I}}) + \mu_3\hat{b}_0\hat{b}_0, \quad \bar{\bar{\varepsilon}} = \varepsilon_0\varepsilon\bar{\bar{I}} \tag{3.10}$$

When $\bar{B}_0 \equiv \hat{b}_0 B_0 = \hat{z}B_0$, i.e., $\hat{b}_0 = \hat{z} = (0, 0, 1)$, the relative permitivity tensor $\bar{\bar{\varepsilon}}$ and the relative permeability tensor $\bar{\bar{\mu}}$ in matrix notation are given by

$$\bar{\bar{\varepsilon}} = \begin{bmatrix} \varepsilon_1 & -i\varepsilon_2 & 0 \\ i\varepsilon_2 & \varepsilon_1 & 0 \\ 0 & 0 & \varepsilon_3 \end{bmatrix} \tag{3.11}$$

and

$$\bar{\bar{\mu}} = \begin{bmatrix} \mu_1 & -i\mu_2 & 0 \\ i\mu_2 & \mu_1 & 0 \\ 0 & 0 & \mu_3 \end{bmatrix} \tag{3.12}$$

It is assumed that the static magnetic field, $\bar{B}_0 = \hat{z}B_0$, is directed in the z direction. For example, if the medium is a cold plasma which is a gyroelectric medium, then the permittivity tensor parameters are given as

$$\varepsilon_1 = 1 - \frac{{\omega_p}^2}{\omega^2 - {\omega_b}^2}, \; \varepsilon_2 = -\frac{\omega_b {\omega_p}^2}{\omega(\omega^2 - {\omega_b}^2)}, \; \varepsilon_3 = 1 - \frac{{\omega_p}^2}{\omega^2} \tag{3.13a}$$

where

$$\omega_b = -\frac{eB_0}{m}, \quad \omega_p = \left(\frac{N_0 e^2}{m\varepsilon_0}\right)^{1/2} \tag{3.13b}$$

ω_b is called the gyrofrequency or cyclotron frequency and ω_p is called the plasma frequency. N_0 shows the number of free electrons per unit volume, m represents the mass of each electron with charge e (a negative number). On the other hand if the medium is a ferrite which is a gyromagnetic medium, then the permeability parameters are defined as

$$\mu_1 = 1 + \frac{\omega_o \omega_m}{\omega_o^2 - \omega^2}, \quad \mu_2 = \frac{\omega \omega_m}{\omega_o^2 - \omega^2}, \quad \mu_3 = 1 \tag{3.14a}$$

and

$$\omega_m = \gamma M_o, \quad \omega_o = \gamma H_o \tag{3.14b}$$

ω_m is defined as the Larmor precessional frequency of the electron in the applied magnetic field H_0 and ω_o is defined as the resonant frequency. M_0 is the saturated magnetization vector and is in the same direction as the applied magnetic field H_0. γ is the *gyromagnetic* ratio and its correct value is given as $\gamma = 2.21 \times 10^5 \left(\frac{rad}{s}\right)\left(\frac{m}{A \bullet turns}\right)$ in [1].

3.3 Dispersion Relations and Wave Matrices

We start our analysis by assuming $e^{-i\omega t}$ dependence. The time-harmonic electromagnetic fields satisfy Maxwell's equations

$$\begin{aligned} \nabla \times \bar{E}_{s,t} &= i\omega\mu_0 \bar{\bar{\mu}} . \bar{H}_{s,t} \\ \nabla \times \bar{H}_{s,t} &= -i\omega\varepsilon_0 \bar{\bar{\varepsilon}} \cdot \bar{E}_{s,t} + \bar{J}_{s,t} \end{aligned} \tag{3.15}$$

We look for the solution of the monochromatic plane wave of the form

$$\bar{E}_{s,t} = \bar{E} e^{i\bar{k}.\bar{r}} \tag{3.16}$$

$$\bar{H}_{s,t} = \bar{H} e^{i\bar{k}.\bar{r}} \tag{3.17}$$

where $\bar{k} = (k_x, k_y, k_z)$ is the propagation vector or the wave vector. In the source free region ($\bar{J}_{s,t} = 0$), we find that

$$\bar{k} \times \bar{E} = \omega \bar{B} = \omega\mu_0 \bar{\bar{\mu}} . \bar{H} \tag{3.18}$$

$$\bar{k} \times \bar{H} = -\omega \bar{D} = \omega\varepsilon_0 \bar{\bar{\varepsilon}} \cdot \bar{E} \tag{3.19}$$

Eliminating $\bar{H}$ from (3.18) and (3.19), we obtain the matrix equation for $\bar{E}$:

$$\bar{\bar{W}}_E . \bar{E} = 0 \tag{3.20}$$

where

$$\bar{\bar{W}}_E = \bar{\bar{k}} \bar{\bar{\mu}}^{-1} \bar{\bar{k}} + k_0^2 \bar{\bar{\varepsilon}} \tag{3.21}$$

where

$$k_0^2 = \omega^2 \mu_0 \varepsilon_0$$

$$\bar{\bar{k}} = \begin{bmatrix} 0 & -k_z & k_y \\ k_z & 0 & -k_x \\ -k_y & k_x & 0 \end{bmatrix}$$

We have used the following identity in (2.8):

$$\bar{k} \times \bar{E} = \bar{\bar{k}} . \bar{E}$$

We define $\bar{\bar{W}}_E$ as an *electric wave matrix*. A non-zero solution $\bar{E}_0$ exists only if the determinant of the electric wave matrix is zero, i.e.,

$$\left| \bar{\bar{W}}_E \right| = \left| \bar{\bar{k}} \bar{\bar{\mu}}^{-1} \bar{\bar{k}} + k_0^2 \bar{\bar{\varepsilon}} \right| = 0 \tag{3.22}$$

The relation in (3.22) is known as the *dispersion relation* for the gyrotropic medium. Equation (3.22) has two positive roots in k^2. They represent two types of waves – type I wave which is represented by the wavenumber k_I, and type II wave which is represented by the wavenumber k_{II}. We assume that the static magnetic field is directed in the z direction, and the angle between the magnetic field and the wave vector is θ. When this angle is $\theta = 0$ or $\frac{\pi}{2}$, the waves propagating inside the medium will be called as *principle waves*. The wave propagation is called longitudinal propagation (with respect to static magnetic field) when $\theta = 0$, and transverse propagation when $\theta = \frac{\pi}{2}$. For the longitudinal propagation, there exist two circularly polarized waves. They are the *right hand circularly polarized wave* and the *left hand circularly polarized wave*. For the transverse propagation, there also exist two waves. They are the *ordinary wave* and the *extraordinary wave*. Since gyroelectric and gyromagnetic media are dual of each other, we choose to give a detailed analysis of wave propagation and dispersion characteristics in a gyroelectric medium in the following sections.

An electron plasma becomes anisotropic under static magnetic field $\bar{B}_0$. If this medium is characterized by a Hermitian permitivity tensor which is given by (3.9) or (3.11), then it is called a gyroelectric or electrically gyrotropic medium. In the plasma, each electron is acted upon by a force $e\bar{E}$ arising from the electric field of the wave and a force $e(\bar{v} \times \bar{B}_0)$ arising from the motion of the electron with average velocity $\bar{v}$ through the constant magnetic field $\bar{B}_0$. For simplicity, we shall ignore the collisions among particles by assuming that the medium is lossless. The equation of motion of electron in this case can be written as

$$- i\omega m\bar{v} = e(\bar{E} + \bar{v} \times \bar{B}_0) \tag{3.23}$$

and yields the following expression for the velocity:

$$\bar{v} = \frac{ie}{m\omega}[\bar{\bar{I}} - i\frac{\omega_b}{\omega}(\hat{b}_0 \times \bar{\bar{I}})]^{-1} \cdot \bar{E} \tag{3.24}$$

where $\hat{b}_0$ is a unit vector in the direction of $\bar{B}_0 = B\hat{b}_0$ and

$$\omega_b = -\frac{eB_0}{m} \tag{3.25}$$

is the gyro-frequency of the electrons. Since the current density is

$$\bar{J} = N_0 e\bar{v} = \bar{\bar{\sigma}} \cdot \bar{E} \tag{3.26}$$

it follows from (3.24) and (3.26) that the conductivity tensor of the plasma is given by

$$\bar{\bar{\sigma}} = i\omega\varepsilon_0 X[\bar{\bar{I}} - iY(\hat{b}_0 \times \bar{\bar{I}}]^{-1} \tag{3.27}$$

where

$$X = \frac{\omega_p{}^2}{\omega^2} \tag{3.28}$$

$$Y = \frac{\omega_b}{\omega} \tag{3.29}$$

and

$$\omega_p = \left(\frac{N_0 e^2}{m\varepsilon_0}\right)^{1/2} \tag{3.30}$$

is the plasma frequency. Since the complex dielectric tensor $\bar{\bar{\varepsilon}}$ is related to the conductivity tensor $\bar{\bar{\sigma}}$ by

$$\bar{\bar{\varepsilon}} = \bar{\bar{I}} + i\frac{1}{\omega\varepsilon_0}\bar{\bar{\sigma}} \tag{3.31}$$

carrying the inverse in (3.27) and substituting the result into (3.31) gives the explicit form of the dielectric tensor of a gyroelectric medium in dyadic form as

$$\bar{\bar{\varepsilon}} = \varepsilon_1(\bar{\bar{I}} - \hat{b}_0\hat{b}_0) + i\varepsilon_2(\hat{b}_0 \times \bar{\bar{I}}) + \varepsilon_3\hat{b}_0\hat{b}_0 \tag{3.32}$$

where

$$\varepsilon_1 = 1 - \frac{X}{1-Y^2} = 1 - \frac{{\omega_p}^2}{\omega^2 - {\omega_b}^2} \tag{3.33}$$

$$\varepsilon_2 = -\frac{XY}{1-Y^2} = -\frac{\omega_b {\omega_p}^2}{\omega(\omega^2 - {\omega_b}^2)} \tag{3.34}$$

$$\varepsilon_3 = 1 - X = 1 - \frac{{\omega_p}^2}{\omega^2} \tag{3.35}$$

When the direction of the constant magnetic field $\bar{B}_0$ is taken parallel to the z-axis in the Cartesian coordinate system, that is, $\hat{b}_0 = (0,0,1)$, then the matrix representation of (3.32) takes the following form

$$\bar{\bar{\varepsilon}} = \begin{bmatrix} \varepsilon_1 & -i\varepsilon_2 & 0 \\ i\varepsilon_2 & \varepsilon_1 & 0 \\ 0 & 0 & \varepsilon_3 \end{bmatrix} \tag{3.36}$$

Throughout the following analysis lossless gyroelectric medium is considered. Hence $\varepsilon_1, \varepsilon_2$ and ε_3 are all real quantities and $\bar{\bar{\varepsilon}}$ is Hermitian.

3.3.1 Dispersion Relations for Gyrotropic Medium

In this section, we will give detailed analysis and derivation of dispersion relations for electrically gyrotropic medium since there is duality between electrically and magnetically gyrotropic media.

The constitutive relations for a homogeneous lossless gyroelectric medium are

$$\bar{D} = \varepsilon_0 \bar{\bar{\varepsilon}} \cdot \bar{E} \tag{3.37}$$

$$\bar{B} = \mu_0 \bar{H} \tag{3.38}$$

where the relative permitivity or dielectric tensor $\bar{\bar{\varepsilon}}$ is given by (3.36). Using Maxwell's equations in the source-free region

$$\begin{aligned} \nabla \times \bar{E}_{s,t} &= i\omega\mu_0 \bar{H}_{s,t} \\ \nabla \times \bar{H}_{s,t} &= -i\omega\varepsilon_0 \bar{\bar{\varepsilon}} \cdot \bar{E}_{s,t} \end{aligned} \tag{3.39}$$

and looking for the solution of the monochromatic plane wave solution of the form given by equations (3.16)–(3.17).

$$\bar{E}_{s,t} = \bar{E}e^{i\bar{k}.\bar{r}} \tag{3.40}$$

$$H_{s,t} = \bar{H}e^{i\bar{k}.\bar{r}} \tag{3.41}$$

we obtain

$$\bar{k} \times \bar{E} = \omega\mu_0\bar{H} \tag{3.42}$$

$$\bar{k} \times \bar{H} = -\omega\bar{D} = \omega\varepsilon_0\bar{\bar{\varepsilon}} \cdot \bar{E} \tag{3.43}$$

Eliminating $\bar{H}$ from (3.42) and (3.43) and solving for $\bar{E}$ gives

$$\left[k_0{}^2\bar{\bar{\varepsilon}} - k^2\bar{\bar{I}} + \bar{k}\bar{k}\right] \cdot \bar{E} = 0 \tag{3.44}$$

where $k^2 = \bar{k}\,.\,\bar{k}$.

We define

$$\bar{\bar{W}}_E = \left[k_0{}^2\bar{\bar{\varepsilon}} - k^2\bar{\bar{I}} + \bar{k}\bar{k}\right] \tag{3.45}$$

as an electric wave matrix. Equation (3.45) is the same as (3.21) when $\bar{\bar{\mu}} = \mu_0\bar{\bar{I}}$. Hence (3.44) can be expressed as

$$\bar{\bar{W}}_E.\bar{E} = 0 \tag{3.46}$$

A non-zero solution $\bar{E}_0$ exists only if the determinant of the wave matrix is zero, i.e.,

$$\left|\bar{\bar{W}}_E\right| = \left|k_0{}^2\bar{\bar{\varepsilon}} - k^2\bar{\bar{I}} + \bar{k}\bar{k}\right| = 0 \tag{3.47}$$

Here the wave vector $\bar{k}$ is defined as

$$\bar{k} = \bar{k}_\rho + \hat{z}k_z \tag{3.48a}$$

$$\bar{k}_\rho = \hat{x}k_x + \hat{y}k_y \tag{3.48b}$$

Then

$$k^2 = k_x^2 + k_y^2 + k_z^2 \tag{3.49}$$

Equation (3.47) defines the dispersion relation for a gyroelectric medium. Equation (3.47) has two roots in k^2. We will find the wave numbers or the roots of (3.47) using two different methods. In the first method, we will represent the dispersion relation in terms of the angle θ which is the angle between the wave vector $\hat{k}$ which represents the direction of the wave normal and the vector $\hat{b}_0$ which represents the direction of the static magnetic field. The two roots k_I and k_{II} represent the wave numbers for the type I and the type II waves. Then, we give the solutions for the wave numbers k_I and k_{II} in term of θ. In the second method, we represent the dispersion relation in terms of k_{zI} and k_{zII} which represent the wave numbers in the z-direction and then give the solutions for the wave numbers k_{zI} and k_{zII} in terms k_ρ which is the transverse component of the wave vector.

3.3.1.1 Method I: Dispersion Relation in Terms of k

Using the following identity,

$$|C| = |A| + \bar{v}.(adjA).\bar{u} \quad if\ C = A + \bar{u}\bar{v}$$

where C and A are matrices and $\bar{u}$ and $\bar{v}$ are vectors. Hence, (3.47) can be expressed as

$$\left|\left({k_0}^2\bar{\bar{\varepsilon}} - k^2\bar{\bar{I}}\right) + \bar{k}\bar{k}\right| = \left|{k_0}^2\bar{\bar{\varepsilon}} - k^2\bar{\bar{I}}\right| + \bar{k}.\, adj\left(k_0^2\bar{\bar{\varepsilon}} - k^2\bar{\bar{I}}\right).\,\bar{k} = 0 \tag{3.50a}$$

Since

$$\begin{aligned}\left|{k_0}^2\bar{\bar{\varepsilon}} - k^2\bar{\bar{I}}\right| &= \left(\hat{k}.\bar{\bar{I}}.\hat{k}\right)\left|{k_0}^2\bar{\bar{\varepsilon}} - k^2\bar{\bar{I}}\right| \\ &= \hat{k}.\left(\left|{k_0}^2\bar{\bar{\varepsilon}} - k^2\bar{\bar{I}}\right|\bar{\bar{I}}\right).\,\hat{k} \\ &= \hat{k}.\left[\left({k_0}^2\bar{\bar{\varepsilon}} - k^2\bar{\bar{I}}\right).adj\left({k_0}^2\bar{\bar{\varepsilon}} - k^2\bar{\bar{I}}\right)\right].\,\hat{k}\end{aligned} \tag{3.50b}$$

So,

$$\left|\left({k_0}^2\bar{\bar{\varepsilon}} - k^2\bar{\bar{I}}\right) + \bar{k}\bar{k}\right| = \hat{k}.\left(k_0^2\bar{\bar{\varepsilon}}\right).\, adj\left(k_0^2\bar{\bar{\varepsilon}} - k^2\bar{\bar{I}}\right).\,\hat{k} = 0 \tag{3.50c}$$

where $\hat{k} = \frac{\bar{k}}{k}$. If we use another identity,

$$adjC = \lambda^2\bar{\bar{I}} + \lambda\left(A_t\bar{\bar{I}} - A\right) + adjA \quad \text{if}\ C = A + \lambda\bar{\bar{I}}$$

then, $adj(k_0^2\bar{\bar{\varepsilon}} - k^2\bar{\bar{I}})$ can be written as

$$adj(k_0^2\bar{\bar{\varepsilon}} - k^2\bar{\bar{I}}) = k^4\bar{\bar{I}} + k_0^2k^2(\bar{\bar{\varepsilon}} - \bar{\bar{\varepsilon}}_t\bar{\bar{I}}) + k_0^4 adj\bar{\bar{\varepsilon}} \tag{3.51}$$

The subscript t stands for the trace of the matrix. When we substitute (3.51) into (3.50), we get

$$k^4(\hat{k}.\bar{\bar{\varepsilon}}.\hat{k}) + k^2\hat{k}.\left[adj\bar{\bar{\varepsilon}} - (adj\bar{\bar{\varepsilon}})_t\bar{\bar{I}}\right].\hat{k} + |\bar{\bar{\varepsilon}}| = 0 \tag{3.52}$$

Performing the tensor-vector operation in (3.52) using (3.36) and

$$|\bar{\bar{\varepsilon}}| = \varepsilon_3\left(\varepsilon_1^2 - \varepsilon_2^2\right) \tag{3.53}$$

$$adj\bar{\bar{\varepsilon}} = \varepsilon_1\varepsilon_2\bar{\bar{I}} - i\varepsilon_2\varepsilon_3\left(\hat{b}_0 \times \bar{\bar{I}}\right) + (\varepsilon_1 - \varepsilon_3)\hat{b}_0\hat{b}_0 \tag{3.54}$$

$$(adj\bar{\bar{\varepsilon}})_t = \varepsilon_1^2 - \varepsilon_2^2 + 2\varepsilon_1\varepsilon_3 \tag{3.55}$$

we obtain

$$\begin{aligned} k^4\left[\varepsilon_1\sin^2\theta + \varepsilon_3\cos^2\theta\right] + k^2k_0^2\left[\left(\varepsilon_2^2 - \varepsilon_1^2 - \varepsilon_1\varepsilon_3\right)\sin^2\theta - 2\varepsilon_1\varepsilon_3\right] \\ + k_0^4\left[\varepsilon_3\left(\varepsilon_1^2 - \varepsilon_2^2\right)\right] = 0 \end{aligned} \tag{3.56}$$

Equation (3.56) has two roots in k^2. The roots for the fourth order equation in (3.56) are

$$\frac{k_I^2}{k_0^2} = \frac{(\varepsilon_1^2 - \varepsilon_2^2)\sin^2(\theta) + \varepsilon_1\varepsilon_3(1+\cos^2(\theta)) + \left[(\varepsilon_1^2 - \varepsilon_2^2 - \varepsilon_1\varepsilon_3)^2\sin^4(\theta) + 4\varepsilon_2^2\varepsilon_3^2\cos^2(\theta)\right]^{1/2}}{2\left[\varepsilon_1\sin^2(\theta) + \varepsilon_3\cos^2(\theta)\right]} \tag{3.57}$$

and

$$\frac{k_{II}^2}{k_0^2} = \frac{(\varepsilon_1^2 - \varepsilon_2^2)\sin^2(\theta) + \varepsilon_1\varepsilon_3(1+\cos^2(\theta)) - \left[(\varepsilon_1^2 - \varepsilon_2^2 - \varepsilon_1\varepsilon_3)^2\sin^4(\theta) + 4\varepsilon_2^2\varepsilon_3^2\cos^2(\theta)\right]^{1/2}}{2\left[\varepsilon_1\sin^2(\theta) + \varepsilon_3\cos^2(\theta)\right]} \tag{3.58}$$

Equations (3.57) and (3.58) represent the two types of waves – type I wave which is represented by k_I, and type II wave which is represented by k_{II}. The dispersion relation in (3.47) then can be put in the following form:

$$|\bar{W}_E| = k_0^2\left[(\varepsilon_1\sin^2\theta + \varepsilon_3\cos^2\theta)(k^2 - k_I^2)(k^2 - k_{II}^2)\right] = 0 \tag{3.59}$$

3.3.1.2 Method II: Dispersion Relation in Terms of k_z

When (3.36) and (3.48) are substituted into (3.47), we get

$$\left|\bar{\bar{W}}_E\right| = \begin{vmatrix} {k_0}^2\varepsilon_1 - {k_y}^2 - {k_z}^2 & k_x k_y - i\varepsilon_2 {k_0}^2 & k_x k_z \\ k_x k_y + i\varepsilon_2 {k_0}^2 & {k_0}^2\varepsilon_1 - {k_x}^2 - {k_z}^2 & k_y k_z \\ k_x k_z & k_y k_z & {k_0}^2\varepsilon_3 - {k_x}^2 - {k_y}^2 \end{vmatrix} = 0 \quad (3.60)$$

Expansion of $\left|\bar{\bar{W}}_E\right|$ leads to the fourth order equation in k_z as follows:

$$\begin{aligned}\left|\bar{\bar{W}}_E\right| &= {k_z}^4{k_0}^2\varepsilon_3 + {k_z}^2\left[{k_0}^2{k_\rho}^2(\varepsilon_1+\varepsilon_3) - 2{k_0}^2\varepsilon_1\varepsilon_3\right] + \\ &\left[{k_0}^6\varepsilon_3({\varepsilon_1}^2 - {\varepsilon_2}^2) - {k_0}^4{k_\rho}^2({\varepsilon_1}^2 - {\varepsilon_2}^2 + \varepsilon_1\varepsilon_3) + {k_0}^2{k_\rho}^4\varepsilon_1\right] = 0\end{aligned} \quad (3.61)$$

This equation has two roots in k_z^2 as

$$\frac{{k_{zI}}^2}{{k_0}^2} = \frac{\left[2\varepsilon_1\varepsilon_3 - \frac{{k_\rho}^2}{{k_0}^2}(\varepsilon_1+\varepsilon_3)\right] + \left[\frac{{k_\rho}^4}{k_0^4}(\varepsilon_1-\varepsilon_3)^2 + 4{\varepsilon_2}^2\varepsilon_3\left(\varepsilon_3 - \frac{{k_\rho}^2}{{k_0}^2}\right)\right]^{1/2}}{2\varepsilon_3} \quad (3.62)$$

and

$$\frac{{k_{zII}}^2}{{k_0}^2} = \frac{\left[2\varepsilon_1\varepsilon_3 - \frac{{k_\rho}^2}{{k_0}^2}(\varepsilon_1+\varepsilon_3)\right] - \left[\frac{{k_\rho}^4}{k_0^4}(\varepsilon_1-\varepsilon_3)^2 + 4{\varepsilon_2}^2\varepsilon_3\left(\varepsilon_3 - \frac{{k_\rho}^2}{{k_0}^2}\right)\right]^{1/2}}{2\varepsilon_3} \quad (3.63)$$

The wavenumbers given in (3.62) and (3.63) correspond to the type I and type II waves, respectively. Now, (3.61) can be expressed as

$$|\bar{\bar{W}}_E| = {k_0}^2\varepsilon_3({k_z}^2 - {k_{zI}}^2)({k_z}^2 - {k_{zII}}^2) = 0 \quad (3.64)$$

We note that when we define the angle between the wave vector $\bar{k}$ and constant magnetic field $\bar{B}_0$ to be θ, one can rewrite the components of the wave vectors in terms of θ as

$$k_\rho = k\sin\theta \quad (3.65)$$

$$k_z = k\cos\theta \quad (3.66)$$

If we substitute (3.65) and (3.66) into the (3.61), we obtain the dispersion relation given by (3.56) which has the solutions for the two types of the wave numbers given by (3.57) and (3.58).

3.4 Plane Waves in Gyrotropic Medium

In this section, we will analyze the problem of finding the polarization of the plane waves which is described by the vector $\bar{E}_0$ in a gyroelectric medium (gyromagnetic medium is dual) such as magnetically biased plasma. The electric field $\bar{E}$ must satisfy the Helmholtz equation

$$\nabla \times \nabla \times \bar{E}_{s,t} = \omega^2 \mu_0 \varepsilon_0 \bar{\bar{\varepsilon}} . \bar{E}_{s,t} \tag{3.67}$$

which is derived from Maxwell's equations

$$\nabla \times \bar{E}_{s,t} = i\omega\mu_0 \bar{H}_{s,t} \tag{3.68}$$

$$\nabla \times \bar{H}_{s,t} = -i\omega\varepsilon_0 \bar{\bar{\varepsilon}} . \bar{E}_{s,t} \tag{3.69}$$

by taking the curl of (3.68) and using (3.69) to eliminate $\bar{H}$. Substituting (3.16) into (3.67), we obtain

$$\bar{E} - \hat{k}(\hat{k} . \bar{E}) = \frac{\omega^2 \mu_0 \varepsilon_0}{k^2} \bar{\bar{\varepsilon}} . \bar{E}$$

or

$$\bar{E} - \hat{k}(\hat{k} . \bar{E}) = \frac{1}{(k/k_0)^2} \bar{\bar{\varepsilon}} . \bar{E} \tag{3.70}$$

If we write the propagation wave vector $\bar{k}$ as

$$\bar{k} = \hat{k} \frac{\omega}{v_p} \tag{3.71}$$

where v_p is the phase velocity of the wave, then we can express (3.70) as

$$\bar{E} - \hat{k}(\hat{k} . \bar{E}) = \frac{v_p^2}{c^2} \bar{\bar{\varepsilon}} . \bar{E} \tag{3.72}$$

where $c = 1/\sqrt{\mu_0 \varepsilon_0}$ is the velocity of light in free space. Without loss of generality, we choose the Cartesian coordinate system such that the z axis is parallel to $\bar{B}_0$ and the yz plane contains $\hat{k}$. As shown in Fig. 3.1, the angle between $\hat{k}$ and $\bar{B}_0$ is denoted by θ.

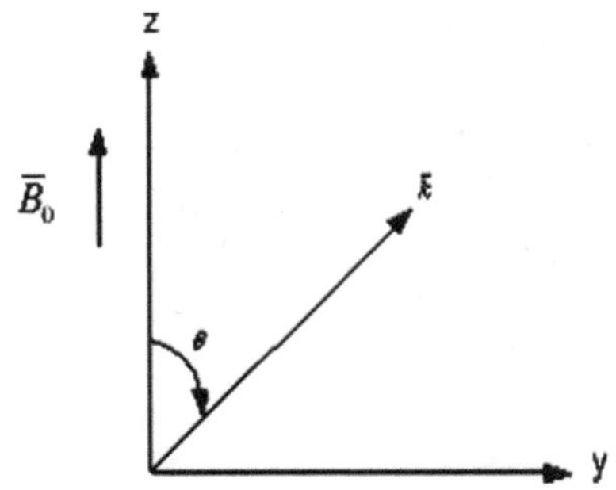

Fig. 3.1 Wave propagation in a gyroelectric medium with an arbitrary direction of $\bar{k}$ and applied external magnetic field $\bar{B}_0$

Accordingly, the x, y, z components of the vector equation (3.72) are given by

$$E_x\left(\frac{k^2}{k_0^2} - \varepsilon_1\right) + i\varepsilon_2 E_y + 0 = 0 \tag{3.73a}$$

$$-i\varepsilon_2 E_x + E_y\left(\frac{k^2}{k_0^2}\cos^2\theta - \varepsilon_1\right) + E_z\left(\frac{k^2}{k_0^2}(-\cos\theta\sin\theta)\right) = 0 \tag{3.73b}$$

$$0 + E_y\left(\frac{k^2}{k_0^2}(-\cos\theta\sin\theta)\right) + E_z\left(\frac{k^2}{k_0^2}\sin^2\theta - \varepsilon_3\right) = 0 \tag{3.73c}$$

where E_x, E_y, E_z are the three components of $\bar{E}$. Since these three simultaneous equations are homogeneous, they yield a nontrivial solution only when

$$\begin{vmatrix} \frac{k^2}{k_0^2} - \varepsilon_1 & i\varepsilon_2 & 0 \\ -i\varepsilon_2 & \frac{k^2}{k_0^2}\cos^2\theta - \varepsilon_1 & -\frac{k^2}{k_0^2}\cos\theta\sin\theta \\ 0 & -\frac{k^2}{k_0^2}\cos\theta\sin\theta & \frac{k^2}{k_0^2}\sin^2\theta - \varepsilon_3 \end{vmatrix} = 0 \tag{3.74}$$

When expanded, (3.74) gives the following relation

$$\tan^2\theta = -\frac{\varepsilon_3\left(\frac{k^2}{k_0^2} - \varepsilon_1 - \varepsilon_2\right)\left(\frac{k^2}{k_0^2} - \varepsilon_1 + \varepsilon_2\right)}{\left(\frac{k^2}{k_0^2} - \varepsilon_3\right)\left(\varepsilon_1\frac{k^2}{k_0^2} - \varepsilon_1^2 + \varepsilon_2^2\right)} \tag{3.75}$$

Equation (3.75) gives the two values of the wave numbers given in (3.57) and (3.58). This is also an alternative method to obtain the wave numbers using Maxwell's equations.

3.4.1 Longitudinal Propagation, $\theta = 0°$

When the direction of phase propagation coincides with the direction of the imposed magnetic field ($\theta = 0°$ or $180°$), we have phenomenon known as *longitudinal propagation*. When the propagation is parallel to $\bar{B}_0$ ($\theta = 0°$), (3.57) and (3.58) reduce to

$$k_I = k_0\sqrt{\varepsilon_1 + \varepsilon_2} \tag{3.76}$$

and

$$k_{II} = k_0\sqrt{\varepsilon_1 - \varepsilon_2} \tag{3.77}$$

Equation (3.73) can be written accordingly as

$$E_x\left(\frac{k^2}{k_0^2} - \varepsilon_1\right) + i\varepsilon_2 E_y = 0 \tag{3.78a}$$

$$-i\varepsilon_2 E_x + E_y\left(\frac{k^2}{k_0^2} - \varepsilon_1\right) = 0 \tag{3.78b}$$

$$E_z(-\varepsilon_3) = 0 \tag{3.78c}$$

From the third of the above equations, we see that E_z is zero. Hence for longitudinal propagation, there is no electric field component in the direction of propagation. Also, it can be shown that the magnetic field $\bar{H}$ is transverse to the direction of propagation. Consequently, the two waves that travel parallel to $\bar{B}_0$ are transverse electromagnetic (TEM) waves. When (3.76) is substituted into (3.78), we obtain

$$\frac{E_x}{E_y} = -i \tag{3.79a}$$

which corresponds to a right-handed circularly polarized (RHCP) wave. If (3.77) is substituted into (3.78), we obtain

$$\frac{E_x}{E_y} = i \tag{3.79b}$$

which corresponds to a left-handed circularly polarized (LHCP) wave. Therefore, the electric field vectors of the two waves traveling parallel to $\bar{B}$ can be written as

$$\bar{E}_I = (\hat{x} + i\hat{y})Ae^{ik_I z} \tag{3.80a}$$

and

$$\bar{E}_{II} = (\hat{x} - i\hat{y})Be^{ik_{II}z} \tag{3.80b}$$

where A and B are arbitrary amplitudes. It is clear that $\bar{E}_I$ represents an RHCP wave, and $\bar{E}_{II}$ represents an LHCP wave. The sum of these two waves yields the following composite wave

$$\bar{E}_I + \bar{E}_{II} = \hat{x}(Be^{ik_{II}z} + Ae^{ik_Iz}) + \hat{y}(-iBe^{ik_{II}z} + iAe^{ik_Iz}) \tag{3.81}$$

To determine the polarization of this composite wave, we consider the ratio $\frac{E_x}{E_y}$. From (3.81), we obtain

$$\frac{E_x}{E_y} = -i\frac{1 + (B/A)e^{i(k_{II}-k_I)z}}{1 - (B/A)e^{i(k_{II}-k_I)z}} \tag{3.82}$$

If the amplitudes of the waves $\bar{E}_I$ and $\bar{E}_{II}$ are chosen to be equal, then the constants A and B become equal. As a result, (3.82) reduces to

$$\frac{E_x}{E_y} = -\cot\left(\frac{k_{II} - k_I}{2}z\right) \quad \text{when } A = B \tag{3.83}$$

Because the ratio in (3.83) is real, the composite wave at any position z is linearly polarized. However, the orientation angle of its plane of polarization (the plane containing $\bar{E}$ and $\bar{k}$) depends on z and rotates as z increases or decreases. In other words, the composite wave undergoes *Faraday rotation*. The angle θ_F through which the resultant vector $\bar{E}$ rotates as the wave travels a unit distance is given by

$$\theta_F = \frac{k_{II} - k_I}{2} \tag{3.84}$$

The rotation is clockwise because always $k_I > k_{II}$. Using (3.76), (3.77) with (3.33) and (3.34), θ_F can be expressed as

$$\theta_F = \frac{k_0}{2}\left[\sqrt{1 - \frac{\omega_p^2}{\omega(\omega + \omega_b)}} - \sqrt{1 - \frac{\omega_p^2}{\omega(\omega - \omega_b)}}\right] \tag{3.85}$$

It is clear that if a wave travels parallel to $\bar{B}_0$ it undergoes a clockwise *Faraday rotation*. On the other hand, if a wave travels anti-parallel to $\bar{B}_0$ it undergoes *Faraday rotation* of the opposite sense. That is, on reversing the direction of propagation, a clockwise wave becomes counterclockwise, and vice versa. This means that if the plane of polarization of a wave traveling parallel to $\bar{B}_0$ is rotated through a certain angle, then upon a reflection it will be rotated still further, the rotation for the round trip being double the rotation of a single crossing. This is one of the properties of non-reciprocal materials such as gyroelectric medium.

3.4.2 Transverse Propagation, $\theta = 90°$

When the direction of wave propagation is perpendicular to the direction of the imposed magnetic field ($\theta = 90°$), we have what is known as *transverse propagation*. When the propagation is perpendicular to $\bar{B}_0$, (3.57) and (3.58) reduce to

$$k_I = k_0\sqrt{\frac{\varepsilon_1^2 - \varepsilon_2^2}{\varepsilon_2}} \tag{3.86}$$

and

$$k_{II} = k_0\sqrt{\varepsilon_3} \tag{3.87}$$

Equation (3.73) can be written accordingly as

$$E_x\left(\frac{k^2}{k_0^2} - \varepsilon_1\right) + i\varepsilon_2 E_y = 0 \tag{3.88a}$$

$$-i\varepsilon_2 E_x + E_y(-\varepsilon_1) = 0 \tag{3.88b}$$

$$E_z\left(\frac{k^2}{k_0^2} - \varepsilon_3\right) = 0 \tag{3.88c}$$

After substituting (3.87) into (3.88), it follows that E_x and E_y are identically zero, and the only surviving component of the electric vector is E_z. Since the propagation constant k_{II} given in (3.87) is independent of $\bar{B}_0$ and equal to the propagation constant of a wave in the isotropic plasma, this TEM wave, known as an *ordinary wave*, is independent of $\bar{B}_0$ in its propagation properties and behaves as it was a TEM wave in isotropic plasma. Thus, we see that one of the two waves traveling in the y direction is a linearly polarized TEM wave whose electric vector is parallel to $\bar{B}_0$ and has the form

$$\bar{E}_{II} = \hat{z}Ae^{ik_{II}y} \tag{3.89}$$

where A is an arbitrary constant. When (3.86) is substituted into (3.88), it is seen that E_z vanishes and that

$$\frac{E_x}{E_y} = i\frac{\varepsilon_1}{\varepsilon_2} \tag{3.90}$$

The electric field vector of this wave, known as an *extraordinary wave*, can be put into the following form

$$\bar{E}_I = \left(i\hat{x}\frac{\varepsilon_1}{\varepsilon_2} + \hat{y}\right)Ce^{ik_I y} \tag{3.91}$$

where C is an arbitrary constant. The magnetic field vector $\bar{H}_I$ is obtained by substituting (3.91) into the (3.69) as

$$\bar{H}_I = -i\hat{z}\frac{k_I}{\omega\mu_0}\frac{\varepsilon_1}{\varepsilon_2}Ce^{ik_I y} \tag{3.92}$$

From (3.91) and (3.92), we see that the extraordinary wave traveling perpendicular to $\bar{B}_0$ is a transverse magnetic (TM) wave with its magnetic vector parallel to $\bar{B}_0$.

Overall, the cases $\theta = 0°$ and $\theta = 90°$, which correspond to longitudinal and transverse propagations, generate two uncoupled waves which are known as *principal waves*.

3.5 Cut-off and Resonance Conditions

The term *cut-off* for any type of the wave occurs when $k = 0$ or the phase velocity, $v_p = \frac{\omega}{k} = \infty$ infinite, and the term *resonance* is used when $k = \infty$ or the phase velocity $v_p = \frac{\omega}{k} = 0$ [2]. *Cut-offs* and *resonances* separate values of the plasma parameters in which the $\frac{k^2}{k_0^2}$ is positive or negative and hence the region of propagation and non-propagation. The attenuation $\sqrt{-k^2}$ is small just beyond cut-off but large just beyond resonance. The characteristics of cut-offs and resonances are listed below in Table 3.1.

The dispersion relation given in (3.56) can be expressed as

$$\left|\bar{\bar{W}}_E\right| = k^4[\varepsilon_1\sin^2(\theta) + \varepsilon_3\cos^2(\theta)] + k_o^2k^2[(\varepsilon_2^2 - \varepsilon_1^2 - \varepsilon_1\varepsilon_3)\sin^2(\theta) - 2\varepsilon_1\varepsilon_3] + k_o^4\varepsilon_3(\varepsilon_1^2 - \varepsilon_2^2) = 0 \tag{3.93}$$

Table 3.1 Characteristics of cut-offs and resonances

Cut-off	Resonance
$v_p = \infty$	$v_p = 0$
$k = 0$	$k = \infty$

or

$$\left|\bar{\bar{W}}_E\right| = Ak^4 + Bk^2 + C = 0 \tag{3.94}$$

where

$$A = \left[\varepsilon_1 \sin^2(\theta) + \varepsilon_3 \cos^2(\theta)\right] \tag{3.95a}$$

$$B = k_o^2\left[(\varepsilon_2^2 - \varepsilon_1^2 - \varepsilon_1\varepsilon_3)\sin^2(\theta) - 2\varepsilon_1\varepsilon_3\right] \tag{3.95b}$$

$$C = k_o^4 \varepsilon_3 (\varepsilon_1^2 - \varepsilon_2^2) \tag{3.95c}$$

As seen from the above relations, if C $=$ 0 and either A $\neq$ 0 or B $\neq$ 0, at least one root of the equation is zero. This represents then the cut-off condition. As C is independent of θ, the cut-off condition does not depend on the direction of propagation.

Similarly, A $=$ 0 represents the resonance condition. This condition is defined by

$$\tan^2(\theta) = -\frac{\varepsilon_3}{\varepsilon_1} \tag{3.96}$$

In contrast to cut-off condition, resonance condition depends on θ.

3.6 Dispersion Curves and Propagation Characteristics

The expressions in (3.57) and (3.58) have always real values.

$$\frac{k_{I,II}^2}{k_0^2} = (\beta - i\alpha)^2 \tag{3.97}$$

When the value in (3.97) is positive, it is equal to β^2; when it is negative it is equal to $-\alpha^2$. Cold plasma is an example of a gyroelectric medium which satisfies the dispersion relation (3.47) with the permittivity tensor given by (3.36). In the magneto-ionic theory it is customary to use notations X and Y which are used to describe the elements of permittivity tensor for the cold plasma in (3.36). They are given by (3.28) and (3.29) as

$$X = \frac{\omega_p^2}{\omega^2}, \qquad Y = \frac{\omega_b}{\omega}$$

where

$$\omega_p = \left(\frac{N_0 e^2}{m\varepsilon_0}\right)^{1/2}, \qquad \omega_b = -\frac{eB_0}{m}$$

So in our analysis, we will use X and Y to describe the dispersion curves in a gyroelectric medium such as cold plasma for three different cases, namely, the isotropic case, the longitudinal propagation and the transverse propagation.

3.6.1 Isotropic Case, No Magnetic Field, $Y = 0$

It is conventional to represent the dispersion of electromagnetic waves by a plot of the propagation constant k against ω as shown in Fig. 3.2. For a field-free plasma, $\bar{B}_0 = 0$, this gives a hyperbola which *cuts off* below the plasma frequency ω_p. Consequently, radio waves of frequency less than the plasma frequency ω_p for the ionosphere are reflected back to earth. This can be illustrated in Fig. 3.3.

When the external applied magnetic field is zero, i.e., $Y = 0$, (3.57) and (3.58) reduce to

$$\frac{k_I^2}{k_0^2} = \frac{k_{II}^2}{k_0^2} = \varepsilon_3 = 1 - X = 1 - \frac{{\omega_p}^2}{\omega^2} \tag{3.98}$$

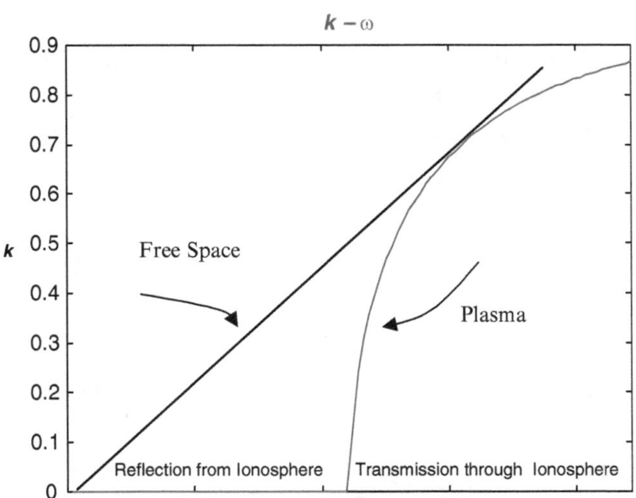

Fig. 3.2 $k - \omega$ diagram for isotropic plasma when $w_p = 1.11 * 10^{12}$

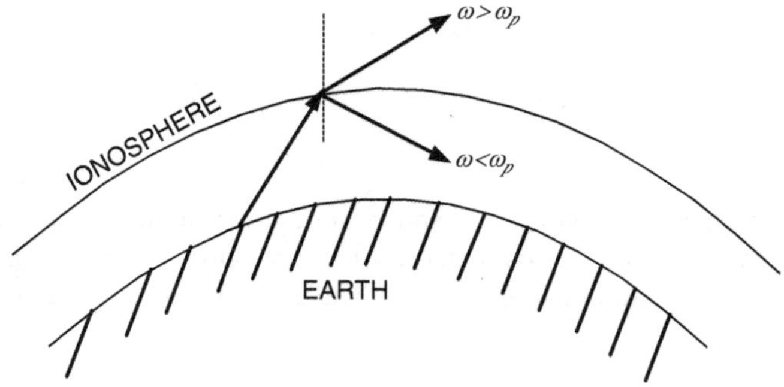

Fig. 3.3 Wave propagation in the ionosphere-earth waveguide

At this stage the plasma is known as *isotropic plasma* because there is no distinction between the type I and type II waves.

3.6.2 The Longitudinal Propagation, $\theta = 0°$

In this section we analyze the cut off and the resonance conditions of the type I and type II waves for the longitudinal propagation. We will use $X - Y^2$ diagram and $k - \omega$ diagram to illustrate the results that we obtain.

3.6.2.1 The Cut-off and Resonance Conditions for Type I Wave

When the cut off condition is met for the type I wave, by setting $k_I = 0$ in (3.76) we obtain the cut off frequency as

$$\omega_{cI_{long}} = \frac{\omega_b}{2} + \sqrt{\frac{\omega_b^2}{4} + \omega_p^2} \qquad (3.99)$$

For waves to propagate, the necessary condition is $k_I^2 > 0$. This requires that

$$\omega > \omega_{cI_{long}} \quad \text{or } Y < 1 - X \qquad (3.100)$$

When the resonance condition is met for type I wave by setting $k_I = \infty$ in (3.76), we get

$$\omega_{rI_{long}} = \omega_b \quad \text{or } Y = 1 \qquad (3.101)$$

The results given by (3.100)–(3.101) can be plotted on the $X - Y^2$ plane as shown in Fig. 3.4.

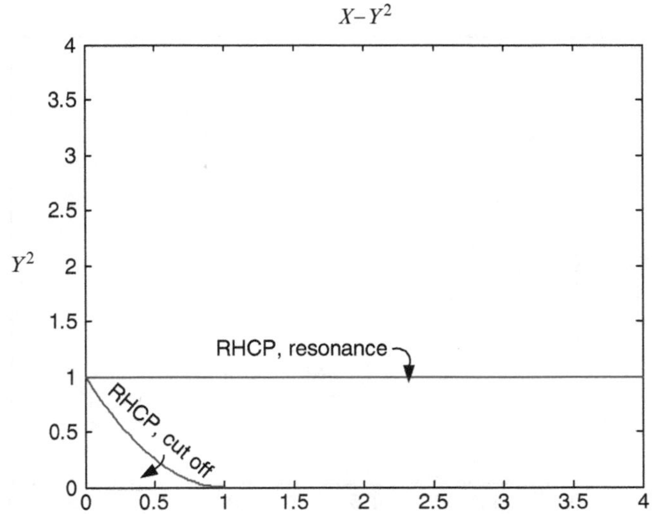

Fig. 3.4 $X - Y^2$ diagram showing the resonance and cut off conditions for longitudinal propagation $\theta = 0°$, for the type I wave

3.6.2.2 The Cut-off and Resonance Conditions for Type II Wave

When the cut off condition is met for the type II wave, from (3.77) we obtain the cut off frequency as

$$\omega_{cII_{long}} = -\frac{\omega_b}{2} + \sqrt{\frac{\omega_b^2}{4} + \omega_p^2} \tag{3.102}$$

For waves to propagate, the necessary condition is $k_{II}^2 > 0$. This requires that

$$\omega > \omega_{cII_{long}} \quad \text{or}\, Y > X - 1 \tag{3.103}$$

No resonance occurs for the type II wave when there is longitudinal propagation. The results can be plotted similarly on the $X - Y^2$ plane as shown in Fig. 3.5.

The same information given by the $X - Y^2$ diagram can be plotted by using the $k - \omega$ diagram shown in Fig. 3.6.

3.6.3 The Transverse Propagation, $\theta = 90°$

In this section we analyze the cut off and the resonance conditions of the type I and type II waves for the transverse propagation. We will use $X - Y^2$ diagram and $k - \omega$ diagram to illustrate the results that we obtain.

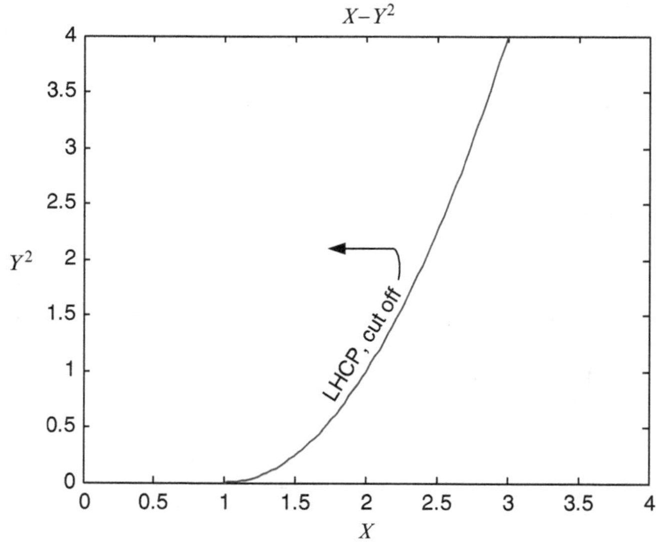

Fig. 3.5 $X - Y^2$ diagram showing the resonance and cut off conditions for longitudinal propagation $\theta = 0°$, for the type II wave

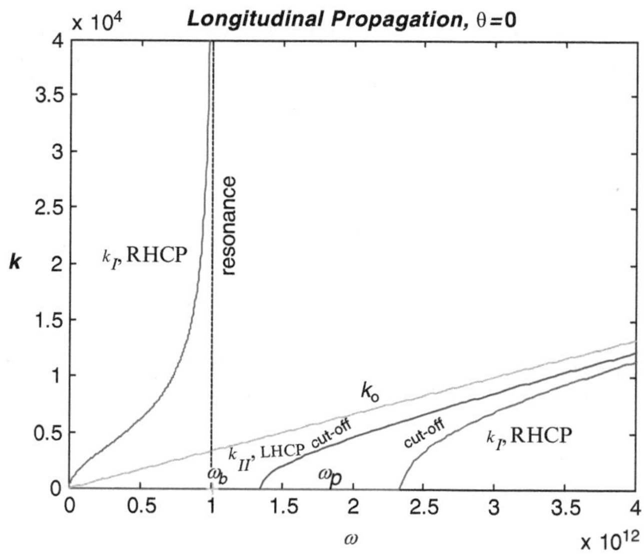

Fig. 3.6 $k - \omega$ diagram showing the resonance and cut off conditions for longitudinal propagation $\theta = 0°$of the type I and type II waves when $\frac{\omega_p^2}{\omega_b^2} = 10^{1/2}$, $w_b = 1 * 10^{12}$

3.6.3.1 The Cut-off and Resonance Conditions for Type I Wave

When the cut off condition is met for the type I wave, from (3.76) we derive the cut off frequency as

$$\omega_{cI_{tran}} = \pm\frac{\omega_b}{2} + \sqrt{\frac{\omega_b^2}{4} + \omega_p^2} \tag{3.104}$$

For waves to propagate, the necessary condition is $k_I^2 > 0$. This requires that

$$\omega > \omega_{cI_{tran}} \tag{3.105}$$

or in terms of X and Y notation

$$Y<1-X \quad \text{when} \quad \omega_{cI_{tran}} = \frac{\omega_b}{2} + \sqrt{\frac{\omega_b^2}{4} + \omega_p^2} \tag{3.106}$$

$$Y>1-X \quad \text{when} \quad \omega_{cI_{tran}} = -\frac{\omega_b}{2} + \sqrt{\frac{\omega_b^2}{4} + \omega_p^2} \tag{3.107}$$

When the resonance condition is met for the type I wave, from (3.76) we get

$$\omega_{rI_{tran}} = \sqrt{\omega_b^2 + \omega_p^2} \quad \text{or} \quad Y^2 = 1 - X \tag{3.108}$$

The results given by (3.104)–(3.108) can be plotted on the $X - Y^2$ plane as shown in Fig. 3.7.

3.6.3.2 The Cut-off and Resonance Conditions for Type II Wave

When the cut off condition is met for type II wave, from (3.77) we obtain the cut off frequency as

$$\omega_{cII_{tran}} = \omega_p \tag{3.109}$$

For waves to propagate, the necessary condition is $k_{II}^2 > 0$. This requires that

$$\omega > \omega_{cII_{tran}} \quad \text{or} \quad X < 1 \tag{3.110}$$

No resonance occurs for type II wave when there is transverse propagation. The results can be plotted similarly on the $X - Y^2$ plane as shown in Fig. 3.8.

The same information given by the $X - Y^2$ diagram can be plotted by using the $k - \omega$ diagram shown in Fig. 3.9.

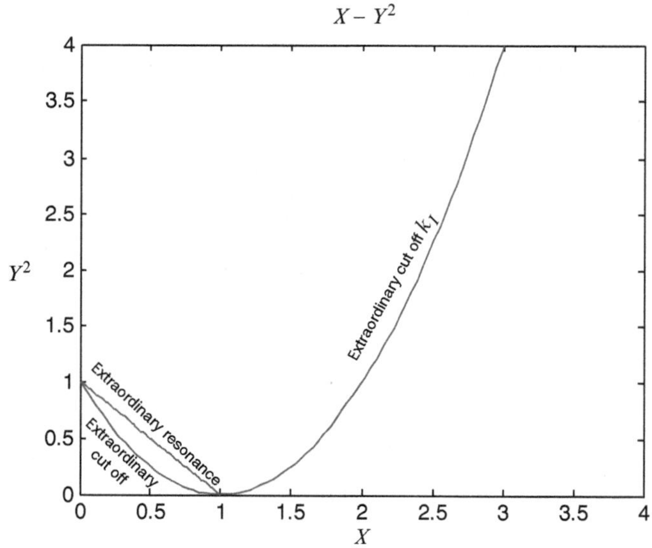

Fig. 3.7 $X - Y^2$ diagram showing the resonance and cut off conditions for longitudinal propagation $\theta = 90°$, for the type I wave

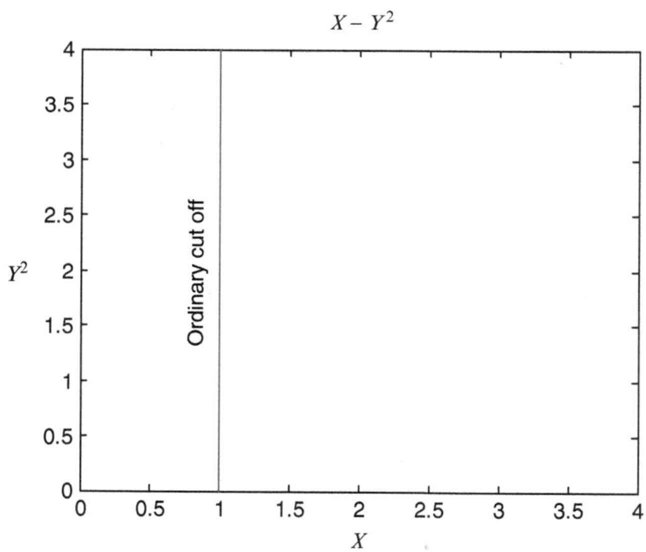

Fig. 3.8 $X - Y^2$ diagram showing the resonance and cut off conditions for longitudinal propagation $\theta = 90°$ for the type II wave

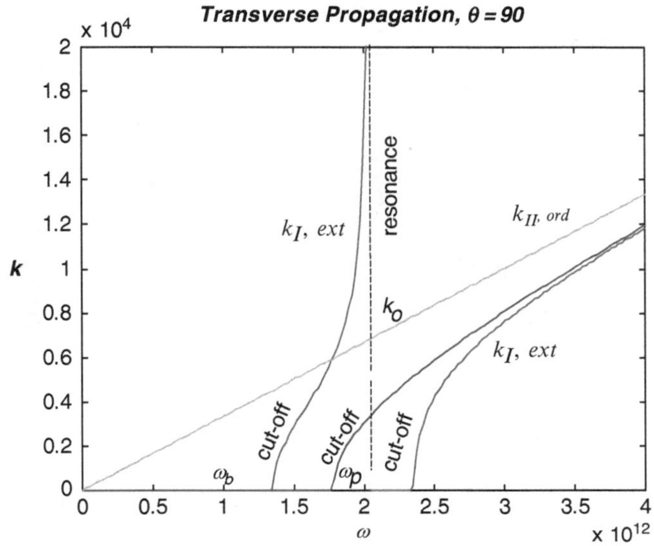

Fig. 3.9 $k - \omega$ diagram showing the resonance and cut off conditions for longitudinal propagation $\theta = 90°$ of the type I and type II waves when $\frac{\omega_p^2}{\omega_b^2} = 10^{1/2}$, $w_b = 1 * 10^{12}$

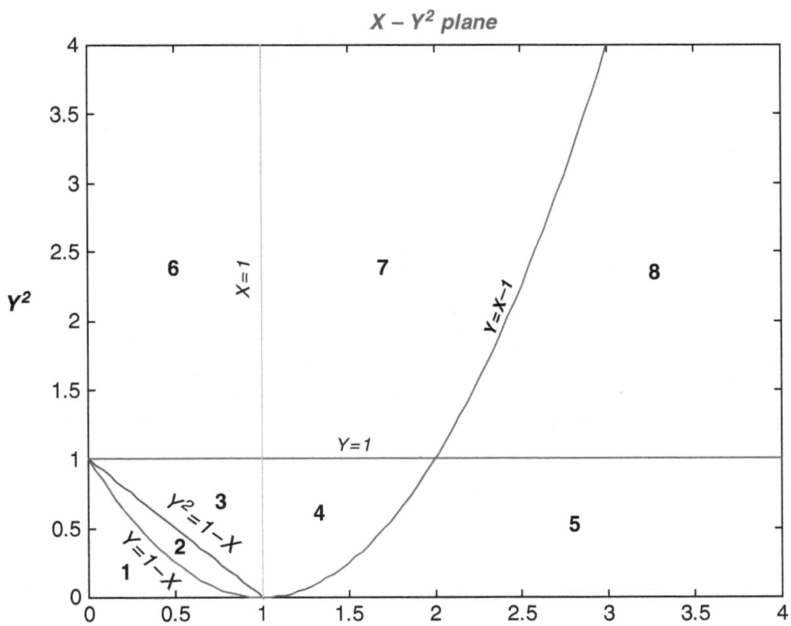

Fig. 3.10 The $X - Y^2$ plane divided into eight regions

Table 3.2 Identification of the frequency bands and the corresponding propagating wave types in each region

	Frequency band	Wave propagates
Region 1	$\omega > \omega_1, \omega_1 = \frac{\omega_b}{2} + \sqrt{\frac{\omega_b^2}{4} + \omega_p^2}$	Type I, type II
Region 2	$\omega_1 > \omega > \omega_2, \omega_2 = \sqrt{\omega_b^2 + \omega_p^2}$	Type I
Region 3	$\omega_2 > \omega > \max(\omega_p, \omega_b)$	Type I, type II
Region 4	$\omega_p > \omega > \max(\omega_b, \omega_3), \omega_3 = \sqrt{\frac{\omega_b^2}{4} + \omega_p^2} - \frac{\omega_b}{2}$	Type II
Region 5	$\omega_3 > \omega > \omega_b$	No propagation
Region 6	$\omega_b > \omega > \omega_p$	Type I, type II
Region 7	$\min(\omega_p, \omega_b) > \omega > \omega_3$	Type I, type II
Region 8	$\min(\omega_3, \omega_b) > \omega > 0$	Type II

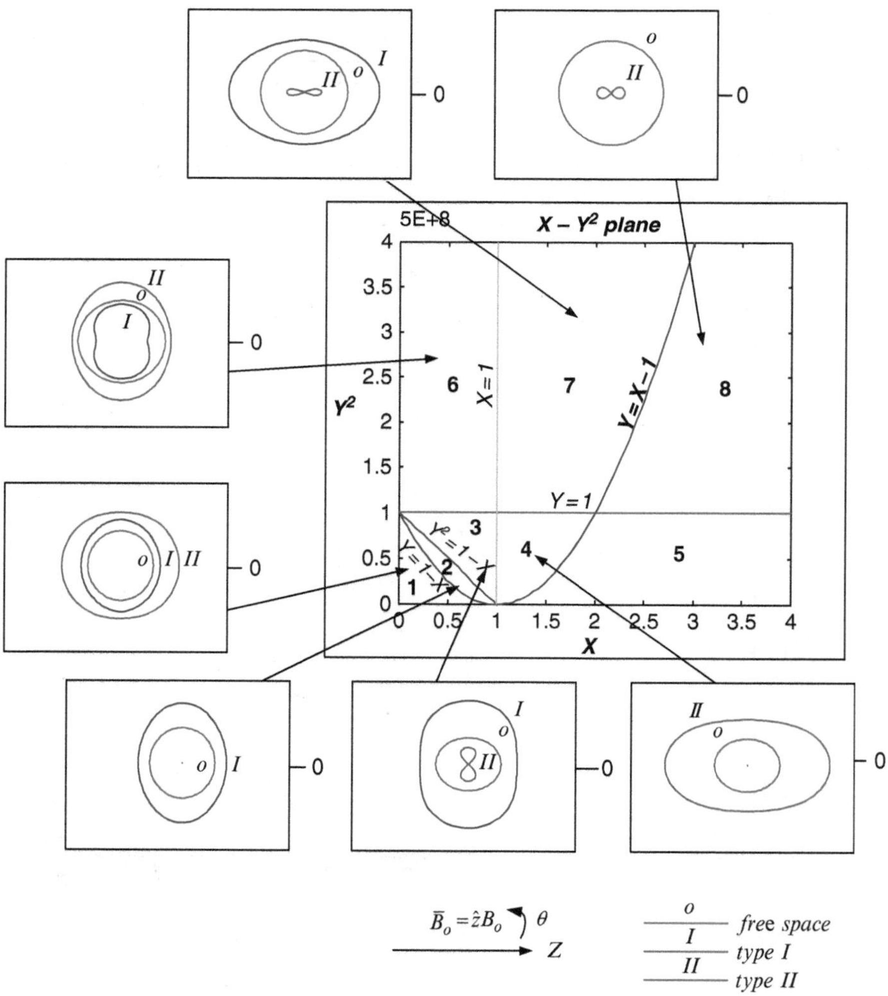

Fig. 3.11 CMA diagram for a cold plasma

3.7 CMA (Clemmow-Mullaly-Allis) Diagram

In Sect. 3.5, we identified the boundaries using principles waves on the $X - Y^2$ plane. Overall, we can divide the $X - Y^2$ plane into eight different regions using the cut off and resonance conditions for the principal waves. When the eight regions given in Sects. 3.5.2 and 3.5.3 are shown in a single diagram, we obtain our base $X - Y^2$ plane as shown in Fig. 3.10.

We obtained the frequency bands for each region over which wave can propagate in Sects. 3.5.2 and 3.5.3. Hence, we not only constructed a single diagram showing boundaries for the cut off and resonance conditions for the principle waves but also identified the frequency bands in each region over which wave can propagate with an arbitrary angle to the magnetic field.

This can be visualized in each region by plotting the wave normal surface which is the polar plot of the phase velocity. The distance from the center to the point on the curves denotes the magnitude of the phase velocity along that direction. When the wave normal surfaces are plotted for each region, we obtain the diagram known as the *CMA diagram*. The results shown in the CMA diagram can be tabulated in Table 3.2 to show the frequency bands and the corresponding waves that can propagate in each region.

As seen from Fig. 3.11 and the results of Table 3.2, both characteristic waves propagate in Regions 1, 3, 6, and 7 while in Regions 4 and 8 only the type II wave propagates and in Region 2 only the type I wave propagates. In Region 5, there is no wave propagation.

References

1. H.C. Chen, Theory of Electromagnetic Waves: Coordinate Free Approach, McGraw Hill, 1983, Chapter 7.
2. W.P. Allis, S.J. Buchsbaum and A. Bers, *Waves in Anisotropic Plasmas,* MIT Press, Cambridge, Massachusetts, 1963.

Chapter 4
Method of Dyadic Green's Functions

4.1 Introduction

In this chapter, method of dyadic Green's function (DGF) for electromagnetic wave problems is introduced. A very practical method to obtain dyadic Green's function for general anisotropic medium is discussed. The duality relations between DGFs are obtained and their application is illustrated. DGFs for unbounded and layered uniaxially anisotropic, unbounded and layered gyrotropic medium are derived.

4.2 Dyadic Green's Functions

In electromagnetic applications such as geophysical prospecting, remote sensing, wave propagation, microstrip circuits and antennas, it is necessary to compute the electromagnetic fields in the medium. Electromagnetic fields can easily be calculated if the Green's function of the medium is known. Derivation of DGFs begins from Maxwell's equations. In the following section, DGFs in differential form will be derived for an unbounded general anisotropic medium illustrated in Fig. 4.1.

4.3 Theory of Dyadic Differential Functions

Maxwell equations for the problem illustrated in Fig. 4.1 in the presence of impressed magnetic current density $\bar{M}$ and the electric current density $\bar{J}$ can be written as

$$\nabla \times \bar{E} = i\omega\mu_0\bar{\bar{\mu}}.\bar{H} - \bar{M} \tag{4.1}$$

$$\nabla \times \bar{H} = -i\omega\varepsilon_0\bar{\bar{\varepsilon}}.\bar{E} + \bar{J} \tag{4.2}$$

A. Eroglu, *Wave Propagation and Radiation in Gyrotropic and Anisotropic Media*,
DOI 10.1007/978-1-4419-6024-5_4, © Springer Science+Business Media, LLC 2010

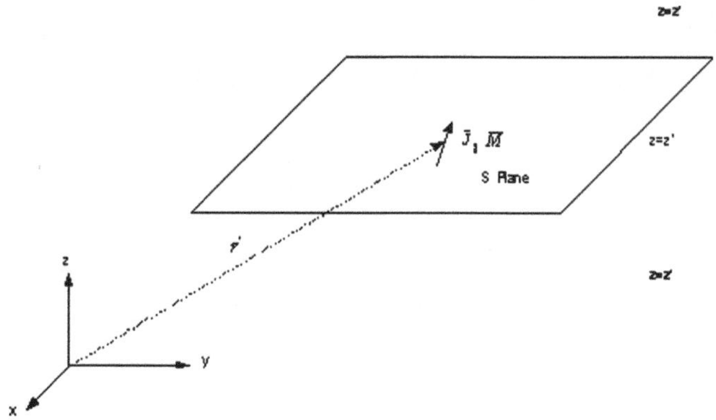

Fig. 4.1 Geometry of an unbounded anisotorpic medium in the existence of current densities

The linearity of Maxwell's equations implies linear dependence of $\bar{E}$ and $\bar{H}$ on the excitations $\bar{J}$ and $\bar{M}$. Then, $\bar{E}$ and $\bar{H}$ in Fig. 4.1 at any point can be represented as

$$\bar{E}(\bar{r}) = \int_{V'} \bar{\bar{G}}_{ee}(\bar{r},\bar{r}').\bar{J}(\bar{r}')d^3\bar{r}' + \int_{V'} \bar{\bar{G}}_{em}(\bar{r},\bar{r}').\bar{M}(\bar{r}')d^3\bar{r}' \tag{4.3}$$

$$\bar{H}(\bar{r}) = \int_{V'} \bar{\bar{G}}_{me}(\bar{r},\bar{r}').\bar{J}(\bar{r}')d^3\bar{r}' + \int_{V'} \bar{\bar{G}}_{mm}(\bar{r},\bar{r}').\bar{M}(\bar{r}')d^3\bar{r}' \tag{4.4}$$

or in matrix form

$$\begin{bmatrix} \bar{E}(\bar{r}) \\ \bar{H}(\bar{r}) \end{bmatrix} = \int_{V'} \begin{bmatrix} \bar{\bar{G}}_{ee}(\bar{r},\bar{r}') & \bar{\bar{G}}_{em}(\bar{r},\bar{r}') \\ \bar{\bar{G}}_{me}(\bar{r},\bar{r}') & \bar{\bar{G}}_{mm}(\bar{r},\bar{r}') \end{bmatrix} \cdot \begin{bmatrix} \bar{J}(\bar{r}') \\ \bar{M}(\bar{r}') \end{bmatrix} d^3\bar{r}' \tag{4.5}$$

where

$$\bar{J}(\bar{r}) = \int_{V'} \delta(\bar{r} - \bar{r}')\bar{\bar{I}}.\bar{J}(\bar{r}')d^3\bar{r}' \tag{4.6}$$

and

$$\bar{M}(\bar{r}) = \int_{V'} \delta(\bar{r} - \bar{r}')\bar{\bar{I}}.\bar{M}(\bar{r}')d^3\bar{r}' \tag{4.7}$$

The dyadic Green's functions $\bar{\bar{G}}_{ee}(\bar{r},\bar{r}')$, $\bar{\bar{G}}_{mm}(\bar{r},\bar{r}')$ are called electric type and magnetic type and $\bar{\bar{G}}_{me}(\bar{r},\bar{r}')$, $\bar{\bar{G}}_{em}(\bar{r},\bar{r}')$ are called magnetic-electric type and electric-magnetic type DGFs for a general anisotropic medium. The first and the second subscripts show the type of the DGF Green's function. The subscript 'e' refers to an electric type and 'm' refers to a magnetic type DGF. Electric type and magnetic-electric type dyadic Green's functions are obtained by assuming $\bar{M} = 0$ and the existence of the electric current density $\bar{J}$. The magnetic type and electric-magnetic type dyadic Green's functions are obtained by assuming $\bar{J} = 0$ and the existence of the magnetic current density $\bar{M}$.

We start our analysis by substituting (4.3), (4.4), (4.6) and (4.7) into (4.1) and assuming $\bar{M} = 0$. We obtain

$$\nabla \times \left[\int_{V'} \bar{\bar{G}}_{ee}(\bar{r}, \bar{r}').\bar{J}(\bar{r}')d^3\bar{r}' \right] = i\omega\mu_0\bar{\bar{\mu}} \cdot \int_{V'} \bar{\bar{G}}_{me}(\bar{r}, \bar{r}').\bar{J}(\bar{r}')d^3\bar{r}'$$

or

$$\nabla \times \bar{\bar{G}}_{ee}(\bar{r}, \bar{r}') = i\omega\mu_0\bar{\bar{\mu}} \cdot \bar{\bar{G}}_{me}(\bar{r}, \bar{r}') \tag{4.8}$$

Substituting (4.3), (4.4), (4.6) and (4.7) into (4.2) and assuming $\bar{M} = 0$ gives

$$\nabla \times \left[\int_{V'} \bar{\bar{G}}_{me}(\bar{r}, \bar{r}').\bar{J}(\bar{r}')d^3\bar{r}' \right] = -i\omega\varepsilon_0\bar{\bar{\varepsilon}}. \int_{V'} \bar{\bar{G}}_{ee}(\bar{r}, \bar{r}').\bar{J}(\bar{r}')d^3\bar{r}' + \int_{V'} \delta(\bar{r} - \bar{r}')\bar{\bar{I}}.\bar{J}(\bar{r}')d^3\bar{r}'$$

or

$$\nabla \times \bar{\bar{G}}_{me}(\bar{r}, \bar{r}') = -i\omega\varepsilon_0\bar{\bar{\varepsilon}}.\bar{\bar{G}}_{ee}(\bar{r}, \bar{r}') + \delta(\bar{r} - \bar{r}')\bar{\bar{I}} \tag{4.9}$$

When we substitute (4.3), (4.4), (4.6) and (4.7) into (4.1) and assume $\bar{J} = 0$, we obtain

$$\nabla \times \left[\int_{V'} \bar{\bar{G}}_{em}(\bar{r}, \bar{r}').\bar{M}(\bar{r}')d^3\bar{r}' \right] = i\omega\mu_0\bar{\bar{\mu}} \cdot \int_{V'} \bar{\bar{G}}_{mm}(\bar{r}, \bar{r}').\bar{M}(\bar{r}')d^3\bar{r}' - \int_{V'} \delta(\bar{r} - \bar{r}')\bar{\bar{I}}.\bar{M}(\bar{r}')d^3\bar{r}'$$

or

$$\nabla \times \bar{\bar{G}}_{em}(\bar{r}, \bar{r}') = i\omega\mu_0\bar{\bar{\mu}} \cdot \bar{\bar{G}}_{mm}(\bar{r}, \bar{r}') - \delta(\bar{r} - \bar{r}')\bar{\bar{I}} \tag{4.10}$$

In the same way, substituting (4.3), (4.4), (4.6) and (4.7) into (4.2) gives

$$\nabla \times \left[\int_{V'} \bar{\bar{G}}_{mm}(\bar{r},\bar{r}').\bar{M}(\bar{r}')d^3\bar{r}' \right] = -i\omega\varepsilon_0\bar{\bar{\varepsilon}}. \int_{V'} \bar{\bar{G}}_{em}(\bar{r},\bar{r}').\bar{M}(\bar{r}')d^3\bar{r}'$$

or

$$\nabla \times \bar{\bar{G}}_{mm}(\bar{r},\bar{r}') = -i\omega\varepsilon_0\bar{\bar{\varepsilon}}.\bar{\bar{G}}_{em}(\bar{r},\bar{r}') \tag{4.11}$$

Equation (4.10) can be re-written as,

$$\bar{\bar{G}}_{mm}(\bar{r},\bar{r}') = \frac{1}{i\omega\mu_0}\bar{\bar{\mu}}^{-1}\cdot\nabla\times\bar{\bar{G}}_{em}(\bar{r},\bar{r}') + \frac{1}{i\omega\mu_0}\bar{\bar{\mu}}^{-1}\delta(\bar{r}-\bar{r}') \tag{4.12}$$

Taking curl of (4.12) gives

$$\nabla\times\bar{\bar{G}}_{mm}(\bar{r},\bar{r}') = \frac{1}{i\omega\mu_0}\nabla\times\bar{\bar{\mu}}^{-1}\cdot\nabla\times\bar{\bar{G}}_{em}(\bar{r},\bar{r}') + \frac{1}{i\omega\mu_0}\nabla\times\bar{\bar{\mu}}^{-1}\delta(\bar{r}-\bar{r}') \tag{4.13}$$

When we substitute (4.11) into (4.13), we get

$$-i\omega\varepsilon_0\bar{\bar{\varepsilon}}.\bar{\bar{G}}_{em}(\bar{r},\bar{r}') = \frac{1}{i\omega\mu_0}\nabla\times\bar{\bar{\mu}}^{-1}\cdot\nabla\times\bar{\bar{G}}_{em}(\bar{r},\bar{r}') + \frac{1}{i\omega\mu_0}\nabla\times\bar{\bar{\mu}}^{-1}\delta(\bar{r}-\bar{r}')$$

or

$$\nabla\times\bar{\bar{\mu}}^{-1}\cdot\nabla\times\bar{\bar{G}}_{em}(\bar{r},\bar{r}') - k_0^2\bar{\bar{\varepsilon}}.\bar{\bar{G}}_{em}(\bar{r},\bar{r}') = -\nabla\times\bar{\bar{\mu}}^{-1}\delta(\bar{r}-\bar{r}') \tag{4.14}$$

Equation (4.14) can be expressed in the following form

$$\left[\nabla\times\bar{\bar{\mu}}^{-1}\cdot\nabla\times\bar{\bar{I}} - k_0^2\bar{\bar{\varepsilon}}\right]\cdot\bar{\bar{G}}_{em}(\bar{r},\bar{r}') = -\nabla\times\bar{\bar{\mu}}^{-1}\delta(\bar{r}-\bar{r}') \tag{4.15}$$

To find the second order dyadic differential equations for magnetic type DGF, we re-write (4.11) as follows

$$\bar{\bar{G}}_{em}(\bar{r},\bar{r}') + \frac{1}{i\omega\varepsilon_0}\bar{\bar{\varepsilon}}^{-1}.\nabla\times\bar{\bar{G}}_{mm}(\bar{r},\bar{r}') = 0 \tag{4.16}$$

Taking curl of (4.16) gives,

$$\nabla\times\bar{\bar{G}}_{em}(\bar{r},\bar{r}') + \frac{1}{i\omega\varepsilon_0}\nabla\times\bar{\bar{\varepsilon}}^{-1}.\nabla\times\bar{\bar{G}}_{mm}(\bar{r},\bar{r}') = 0$$

or

$$\nabla \times \bar{\bar{G}}_{em}(\bar{r},\bar{r}') = -\frac{1}{i\omega\varepsilon_0}\nabla \times \bar{\bar{\varepsilon}}^{-1}.\nabla \times \bar{\bar{G}}_{mm}(\bar{r},\bar{r}') \tag{4.17}$$

When we substitute (4.10) into (4.17), we get the second order dyadic differential equation for the electric-magnetic type DGF function as

$$\frac{1}{i\omega\varepsilon_0}\nabla \times \bar{\bar{\varepsilon}}^{-1}.\nabla \times \bar{\bar{G}}_{mm}(\bar{r},\bar{r}') + i\omega\mu_0\bar{\bar{\mu}} \cdot \bar{\bar{G}}_{mm}(\bar{r},\bar{r}') = \bar{\bar{I}}\delta(\bar{r}-\bar{r}')$$

or

$$\nabla \times \bar{\bar{\varepsilon}}^{-1}.\nabla \times \bar{\bar{G}}_{mm}(\bar{r},\bar{r}') - k_0^2\bar{\bar{\mu}} \cdot \bar{\bar{G}}_{mm}(\bar{r},\bar{r}') = i\omega\varepsilon_0\bar{\bar{I}}\delta(\bar{r}-\bar{r}')$$

which can be rewritten as

$$\left[\nabla \times \bar{\bar{\varepsilon}}^{-1}.\nabla \times \bar{\bar{I}} - k_0^2\bar{\bar{\mu}}\right].\bar{\bar{G}}_{mm}(\bar{r},\bar{r}') = i\omega\varepsilon_0\bar{\bar{I}}\delta(\bar{r}-\bar{r}') \tag{4.18}$$

We can obtain the second order dyadic differential equation for electric type dyadic Green's function by re-expressing (4.8) as

$$\bar{\bar{G}}_{me}(\bar{r},\bar{r}') = \frac{1}{i\omega\mu_0}\bar{\bar{\mu}}^{-1} \cdot \nabla \times \bar{\bar{G}}_{ee}(\bar{r},\bar{r}') \tag{4.19}$$

Take the curl of (4.19) and obtain

$$\nabla \times \bar{\bar{G}}_{me}(\bar{r},\bar{r}') = \frac{1}{i\omega\mu_0}\nabla \times \bar{\bar{\mu}}^{-1} \cdot \nabla \times \bar{\bar{G}}_{ee}(\bar{r},\bar{r}') \tag{4.20}$$

Substituting (4.20) into (4.9) gives

$$\frac{1}{i\omega\mu_0} \cdot \nabla \times \bar{\bar{\mu}}^{-1} \cdot \nabla \times \bar{\bar{G}}_{ee}(\bar{r},\bar{r}') + i\omega\varepsilon_0\bar{\bar{\varepsilon}}.\bar{\bar{G}}_{ee}(\bar{r},\bar{r}') = \delta(\bar{r}-\bar{r}')\bar{\bar{I}}$$

or

$$\nabla \times \bar{\bar{\mu}}^{-1} \cdot \nabla \times \bar{\bar{G}}_{ee}(\bar{r},\bar{r}') - k_0^2\bar{\bar{\varepsilon}}.\bar{\bar{G}}_{ee}(\bar{r},\bar{r}') = i\omega\mu_0\delta(\bar{r}-\bar{r}')\bar{\bar{I}}$$

which can be rewritten as

$$\left[\nabla \times \bar{\bar{\mu}}^{-1}.\nabla \times \bar{\bar{I}} - k_0^2\bar{\bar{\varepsilon}}\right].\bar{\bar{G}}_{ee}(\bar{r},\bar{r}') = i\omega\mu_0\bar{\bar{I}}\delta(\bar{r}-\bar{r}') \tag{4.21}$$

To find the magnetic-electric type DGF, we rewrite the (4.9) as

$$\bar{\bar{G}}_{ee}(\bar{r},\bar{r}')+\frac{1}{i\omega\varepsilon_0}\bar{\bar{\varepsilon}}^{-1}.\nabla\times\bar{\bar{G}}_{me}(\bar{r},\bar{r}')=\frac{1}{i\omega\varepsilon_0}\bar{\bar{\varepsilon}}^{-1}.\delta(\bar{r}-\bar{r}')$$

or

$$\bar{\bar{G}}_{ee}(\bar{r},\bar{r}')=-\frac{1}{i\omega\varepsilon_0}\bar{\bar{\varepsilon}}^{-1}.\nabla\times\bar{\bar{G}}_{me}(\bar{r},\bar{r}')+\frac{1}{i\omega\varepsilon_0}\bar{\bar{\varepsilon}}^{-1}\delta(\bar{r}-\bar{r}') \tag{4.22}$$

Taking the curl of (4.22) gives,

$$\nabla\times\bar{\bar{G}}_{ee}(\bar{r},\bar{r}')=-\frac{1}{i\omega\varepsilon_0}\nabla\times\bar{\bar{\varepsilon}}^{-1}.\nabla\times\bar{\bar{G}}_{me}(\bar{r},\bar{r}')+\frac{1}{i\omega\varepsilon_0}\nabla\times\bar{\bar{\varepsilon}}^{-1}\delta(\bar{r}-\bar{r}') \tag{4.23}$$

Substituting (4.8) into (4.23) gives the second order dyadic differential equation for the magnetic-electric type DGF function as

$$-\frac{1}{i\omega\varepsilon_0}\nabla\times\bar{\bar{\varepsilon}}^{-1}.\nabla\times\bar{\bar{G}}_{me}(\bar{r},\bar{r}')+\frac{1}{i\omega\varepsilon_0}\nabla\times\bar{\bar{\varepsilon}}^{-1}\delta(\bar{r}-\bar{r}')-i\omega\mu_0\bar{\bar{\mu}}\cdot\bar{\bar{G}}_{me}(\bar{r},\bar{r}')=0$$

or

$$\nabla\times\bar{\bar{\varepsilon}}^{-1}.\nabla\times\bar{\bar{G}}_{me}(\bar{r},\bar{r}')-k_0^2\bar{\bar{\mu}}\cdot\bar{\bar{G}}_{me}(\bar{r},\bar{r}')=\nabla\times\bar{\bar{\varepsilon}}^{-1}\delta(\bar{r}-\bar{r}')$$

which can be rewritten as

$$\left[\nabla\times\bar{\bar{\varepsilon}}^{-1}.\nabla\times\bar{\bar{I}}-k_0^2\bar{\bar{\mu}}\right].\bar{\bar{G}}_{me}(\bar{r},\bar{r}')=\nabla\times\bar{\bar{\varepsilon}}^{-1}\delta(\bar{r}-\bar{r}') \tag{4.24}$$

We can summarize the results of the complete set of dyadic first order and second order differential equations for any type of anisotropic medium as follows.

The complete set of the first order dyadic differential equations are,

$$\nabla\times\bar{\bar{G}}_{ee}(\bar{r},\bar{r}')=i\omega\mu_0\bar{\bar{\mu}}\cdot\bar{\bar{G}}_{me}(\bar{r},\bar{r}') \tag{4.25}$$

$$\nabla\times\bar{\bar{G}}_{me}(\bar{r},\bar{r}')=-i\omega\varepsilon_0\bar{\bar{\varepsilon}}.\bar{\bar{G}}_{ee}(\bar{r},\bar{r}')+\delta(\bar{r}-\bar{r}')\bar{\bar{I}} \tag{4.26}$$

$$\nabla\times\bar{\bar{G}}_{mm}(\bar{r},\bar{r}')=-i\omega\varepsilon_0\bar{\bar{\varepsilon}}.\bar{\bar{G}}_{em}(\bar{r},\bar{r}') \tag{4.27}$$

$$\nabla\times\bar{\bar{G}}_{em}(\bar{r},\bar{r}')=i\omega\mu_0\bar{\bar{\mu}}\cdot\bar{\bar{G}}_{mm}(\bar{r},\bar{r}')-\delta(\bar{r}-\bar{r}')\bar{\bar{I}} \tag{4.28}$$

The complete set of the second order dyadic differential equations are,

$$\left[\nabla \times \bar{\bar{\mu}}^{-1}.\nabla \times \bar{\bar{I}} - k_0^2\bar{\bar{\varepsilon}}\right].\bar{\bar{G}}_{ee}(\bar{r},\bar{r}') = i\omega\mu_0\bar{\bar{I}}\delta(\bar{r}-\bar{r}') \tag{4.29}$$

$$\left[\nabla \times \bar{\bar{\varepsilon}}^{-1}.\nabla \times \bar{\bar{I}} - k_0^2\bar{\bar{\mu}}\right].\bar{\bar{G}}_{mm}(\bar{r},\bar{r}') = i\omega\varepsilon_0\bar{\bar{I}}\delta(\bar{r}-\bar{r}') \tag{4.30}$$

$$\left[\nabla \times \bar{\bar{\mu}}^{-1}\cdot\nabla \times \bar{\bar{I}} - k_0^2\bar{\bar{\varepsilon}}\right]\cdot\bar{\bar{G}}_{em}(\bar{r},\bar{r}') = -\nabla \times \bar{\bar{\mu}}^{-1}\delta(\bar{r}-\bar{r}') \tag{4.31}$$

$$\left[\nabla \times \bar{\bar{\varepsilon}}^{-1}.\nabla \times \bar{\bar{I}} - k_0^2\bar{\bar{\mu}}\right].\bar{\bar{G}}_{me}(\bar{r},\bar{r}') = \nabla \times \bar{\bar{\varepsilon}}^{-1}\delta(\bar{r}-\bar{r}') \tag{4.32}$$

4.4 Duality Principle for Dyadic Green's Functions

The symmetry that exists between electric and magnetic fields is also seen on the dyadic Green's functions that are obtained in the preceding section. We write Maxwell's equations below again to observe the symmetry between electric field, $\bar{E}$, and magnetic field, $\bar{H}$.

$$\nabla \times \bar{E} = -\frac{\partial(\bar{\bar{\mu}}\cdot\bar{H})}{\partial t} - \bar{M} \tag{4.33}$$

$$\nabla \times \bar{H} = \frac{\partial(\bar{\bar{\varepsilon}}\cdot\bar{E})}{\partial t} + \bar{J} \tag{4.34}$$

$$\nabla\cdot\bar{B} = \rho_m \tag{4.35}$$

$$\nabla\cdot\bar{D} = \rho_e \tag{4.36}$$

We introduced magnetic charge density ρ_m to illustrate mathematical symmetry although it does not exist in physical world. We write the following replacements to see the symmetry.

$$\bar{E} \to \bar{H}, \quad \bar{\bar{\mu}} \to \bar{\bar{\varepsilon}}, \quad \bar{M} \to -\bar{J} \tag{4.37}$$

$$\bar{H} \to -\bar{E}, \quad \bar{\bar{\varepsilon}} \to \bar{\bar{\mu}}, \quad \bar{J} \to -\bar{M} \tag{4.38}$$

$$\bar{B} \to -\bar{D}, \quad \rho_m \to \rho_e \tag{4.39}$$

$$\bar{D} \to \bar{B}, \quad \rho_e \to \rho_m \tag{4.40}$$

When (4.37) is applied to (4.33), we obtain (4.34), Ampere's law. When the replacement in (4.38) is applied to (4.34), we obtain (4.33), Faraday's law. Similarly, (4.39) and (4.40) can be applied to (4.35) and (4.36) to observe the existing symmetry between Gauss' law for magnetic and electric fields. If we study (4.25)–(4.32), the first order dyadic differential equations and the second order dyadic differential equations, the application of the duality transformation

$$\begin{aligned} \bar{\bar{\varepsilon}} \to \bar{\bar{\mu}}, \quad \bar{\bar{\mu}} \to \bar{\bar{\varepsilon}} \\ \varepsilon_0 \to \mu_0, \quad \mu_0 \to \varepsilon_0 \end{aligned} \tag{4.41}$$

on the dyadic differential equations (4.25)–(4.32) transform one to its dual with the replacements of

$$\bar{\bar{G}}_{ee}(\bar{r},\bar{r}') \to \bar{\bar{G}}_{mm}(\bar{r},\bar{r}') \tag{4.42}$$

$$\bar{\bar{G}}_{me}(\bar{r},\bar{r}') \to -\bar{\bar{G}}_{em}(\bar{r},\bar{r}') \tag{4.43}$$

$$\bar{\bar{G}}_{mm}(\bar{r},\bar{r}') \to \bar{\bar{G}}_{ee}(\bar{r},\bar{r}') \tag{4.44}$$

$$\bar{\bar{G}}_{em}(\bar{r},\bar{r}') \to -\bar{\bar{G}}_{me}(\bar{r},\bar{r}') \tag{4.45}$$

4.5 Formulation of Dyadic Green's Functions

In this section, we'd like to formulate dyadic Green's function in a general form that can be used for any type of unbounded anisotropic medium. We start our analysis by re-writing (4.5) as

$$\begin{bmatrix} \bar{E}(\bar{r}) \\ \bar{H}(\bar{r}) \end{bmatrix} = \int_{V'} \bar{\bar{G}}(\bar{r},\bar{r}') . \begin{bmatrix} \bar{J}(\bar{r}') \\ \bar{M}(\bar{r}') \end{bmatrix} d^3\bar{r}' \tag{4.46}$$

where

$$\bar{\bar{G}}(\bar{r},\bar{r}') = \begin{bmatrix} \bar{\bar{G}}_{ee}(\bar{r},\bar{r}') & \bar{\bar{G}}_{em}(\bar{r},\bar{r}') \\ \bar{\bar{G}}_{me}(\bar{r},\bar{r}') & \bar{\bar{G}}_{mm}(\bar{r},\bar{r}') \end{bmatrix} \tag{4.47}$$

In (4.46), the integral is taken over all space, i.e.

$$\int d^3\bar{r}' = \iint \int_{-\infty}^{\infty} dx' dy' dz' \tag{4.48}$$

The Fourier transform pair of the field vectors can be expressed as

$$\bar{F}(\bar{r}) = \frac{1}{(2\pi)^3}\int \bar{F}(\bar{k})e^{i\bar{k}.\bar{r}}d^3\bar{k} \tag{4.49a}$$

$$\bar{F}(\bar{k}) = \int \bar{F}(\bar{r})e^{-i\bar{k}.\bar{r}}d^3\bar{r} \tag{4.49b}$$

where $\bar{F} = \bar{E}$ for an electric field vector and $\bar{F} = \bar{H}$ for a magnetic field vector. Here the integration over $\bar{k}$ is three dimensional as the integration over $\bar{r}$, i.e., $d^3\bar{k} = dk_x dk_y dk_z$.

Similarly, the Fourier transform pairs for DGF $\bar{\bar{G}}(\bar{r},\bar{r}')$ can be written as

$$\bar{\bar{G}}(\bar{r},\bar{r}') = \frac{1}{(2\pi)^3}\int \bar{\bar{G}}(\bar{k})e^{i\bar{k}.(\bar{r}-\bar{r}')}d^3\bar{k} \tag{4.50}$$

$$\bar{\bar{G}}(\bar{k}) = \int \bar{\bar{G}}(\bar{r},\bar{r}')e^{-i\bar{k}.(\bar{r}-\bar{r}')}d^3(\bar{r}-\bar{r}') \tag{4.51}$$

Translational invariance assumed in the use of $(\bar{r}-\bar{r}')$ comes from the unbounded nature of the problem. Substitution of (4.49a), (4.50) into (4.46) gives

$$\frac{1}{(2\pi)^3}\int \begin{bmatrix} \bar{E}(\bar{k}) \\ \bar{H}(\bar{k}) \end{bmatrix} e^{i\bar{k}.\bar{r}}d^3\bar{k} = \int \left[\frac{1}{(2\pi)^3}\int \bar{\bar{G}}(\bar{k})e^{i\bar{k}.(\bar{r}-\bar{r}')}d^3\bar{k}\right] \begin{bmatrix} \bar{J}(\bar{r}') \\ \bar{M}(\bar{r}') \end{bmatrix} d^3\bar{r}' \tag{4.52}$$

where $\bar{\bar{G}}(\bar{k})$ is given by

$$\bar{\bar{G}}(\bar{k}) = \begin{bmatrix} \bar{\bar{G}}_{ee}(\bar{k}) & \bar{\bar{G}}_{em}(\bar{k}) \\ \bar{\bar{G}}_{me}(\bar{k}) & \bar{\bar{G}}_{mm}(\bar{k}) \end{bmatrix} \tag{4.53}$$

When we change the order of integration in (4.52), we obtain

$$\begin{aligned} \int \begin{bmatrix} \bar{E}(\bar{k}) \\ \bar{H}(\bar{k}) \end{bmatrix} e^{i\bar{k}.\bar{r}}d^3\bar{k} &= \int \bar{\bar{G}}(\bar{k})e^{i\bar{k}.\bar{r}} \int \begin{bmatrix} \bar{J}(\bar{r}') \\ \bar{M}(\bar{r}') \end{bmatrix} e^{-i\bar{k}.\bar{r}'}d^3\bar{r}'d^3\bar{k} \\ &= \int \bar{\bar{G}}(\bar{k}) \begin{bmatrix} \bar{J}(\bar{k}) \\ \bar{M}(\bar{k}) \end{bmatrix} e^{i\bar{k}.\bar{r}}d^3\bar{k} \end{aligned} \tag{4.54}$$

Using (4.54), we can relate the field vectors to dyadic Green's functions in the k-domain as

$$\begin{bmatrix} \bar{E}(\bar{k}) \\ \bar{H}(\bar{k}) \end{bmatrix} = \begin{bmatrix} \bar{\bar{G}}_{ee}(\bar{k}) & \bar{\bar{G}}_{em}(\bar{k}) \\ \bar{\bar{G}}_{me}(\bar{k}) & \bar{\bar{G}}_{mm}(\bar{k}) \end{bmatrix} \begin{bmatrix} \bar{J}(\bar{k}) \\ \bar{M}(\bar{k}) \end{bmatrix} \tag{4.55}$$

Now, we assume the solutions for the fields in the form of $e^{i\bar{k}\cdot\bar{r}}$ and use

$$\bar{k}\times\bar{E}=\bar{\bar{k}}\cdot\bar{E} \tag{4.56}$$

where

$$\bar{\bar{k}}=\begin{bmatrix} 0 & -k_z & k_y \\ k_z & 0 & -k_x \\ -k_y & k_x & 0 \end{bmatrix} \tag{4.57}$$

Maxwell's equations given by (4.1) and (4.2) can be transformed into the k-domain and we can derive the matrix equations for $\bar{E}(\bar{k})$ and $\bar{H}(\bar{k})$ as

$$\left[\bar{\bar{k}}\bar{\bar{\varepsilon}}^{-1}\bar{\bar{k}}+k_0^2\bar{\bar{\mu}}\right].\bar{H}(\bar{k})=-i\bar{\bar{k}}\bar{\bar{\varepsilon}}^{-1}\bar{J}(\bar{k})-i\omega\varepsilon_0\bar{M}(\bar{k}) \tag{4.58}$$

$$\left[\bar{\bar{k}}\bar{\bar{\mu}}^{-1}\bar{\bar{k}}+k_0^2\bar{\bar{\varepsilon}}\right].\bar{E}(\bar{k})=-i\omega\mu_0\bar{J}(\bar{k})+i\bar{\bar{k}}\bar{\bar{\mu}}^{-1}\bar{M}(\bar{k}) \tag{4.59}$$

where $k_0^2=\omega^2\mu_0\varepsilon_0$

$$\bar{\bar{W}}_H=\left[\bar{\bar{k}}\bar{\bar{\varepsilon}}^{-1}\bar{\bar{k}}+k_0^2\bar{\bar{\mu}}\right] \tag{4.60}$$

is *magnetic wave matrix* and

$$\bar{\bar{W}}_E=\left[\bar{\bar{k}}\bar{\bar{\mu}}^{-1}\bar{\bar{k}}+k_0^2\bar{\bar{\varepsilon}}\right] \tag{4.61}$$

is *electric wave matrix*. Equations (4.58) and (4.59) can be expressed using wave matrices as

$$\bar{\bar{W}}_H.\bar{H}(\bar{k})=-i\bar{\bar{k}}\bar{\bar{\varepsilon}}^{-1}\bar{J}(\bar{k})-i\omega\varepsilon_0\bar{M}(\bar{k}) \tag{4.62}$$

$$\bar{\bar{W}}_E.\bar{E}(\bar{k})=i\bar{\bar{k}}\bar{\bar{\mu}}^{-1}\bar{M}(\bar{k})-i\omega\mu_0\bar{J}(\bar{k}) \tag{4.63}$$

We can represent (4.62) and (4.63) in matrix form as

$$\begin{bmatrix} \bar{\bar{W}}_E(\bar{k}) & 0 \\ 0 & \bar{\bar{W}}_H(\bar{k}) \end{bmatrix}\begin{bmatrix} \bar{E}(\bar{k}) \\ \bar{H}(\bar{k}) \end{bmatrix}=\begin{bmatrix} -i\omega\mu_0\bar{\bar{I}} & i\bar{\bar{k}}\bar{\bar{\mu}}^{-1} \\ -i\bar{\bar{k}}\bar{\bar{\varepsilon}}^{-1} & -i\omega\varepsilon_0\bar{\bar{I}} \end{bmatrix}\begin{bmatrix} \bar{J}(\bar{k}) \\ \bar{M}(\bar{k}) \end{bmatrix} \tag{4.64}$$

Equation (4.64) can be modified as

$$\begin{bmatrix} \bar{E}(\bar{k}) \\ \bar{H}(\bar{k}) \end{bmatrix}=\begin{bmatrix} -i\omega\mu_0\bar{\bar{W}}_E^{-1} & i\bar{\bar{W}}_E^{-1}\bar{\bar{k}}\bar{\bar{\mu}}^{-1} \\ -i\bar{\bar{W}}_H^{-1}\bar{\bar{k}}\bar{\bar{\varepsilon}}^{-1} & -i\omega\varepsilon_0\bar{\bar{W}}_H^{-1} \end{bmatrix}\begin{bmatrix} \bar{J}(\bar{k}) \\ \bar{M}(\bar{k}) \end{bmatrix} \tag{4.65}$$

Equation (4.65) relates the field vectors to the inverses of the wave matrices in the k-domain. When (4.55) and (4.65) are compared, we obtain the following relation

$$\begin{bmatrix} \bar{\bar{G}}_{ee}(\bar{k}) & \bar{\bar{G}}_{em}(\bar{k}) \\ \bar{\bar{G}}_{me}(\bar{k}) & \bar{\bar{G}}_{mm}(\bar{k}) \end{bmatrix} = \begin{bmatrix} -i\omega\mu_0\bar{\bar{W}}_E^{-1} & i\bar{\bar{W}}_E^{-1}\bar{\bar{k}}\bar{\bar{\mu}}^{-1} \\ -i\bar{\bar{W}}_H^{-1}\bar{\bar{k}}\bar{\bar{\varepsilon}}^{-1} & -i\omega\varepsilon_0\bar{\bar{W}}_H^{-1} \end{bmatrix} \tag{4.66}$$

Equation (4.66) is the representation of DGFs in terms of the inverses of the wave matrices in the k-domain for a general anisotropic medium. More explicitly,

$$\bar{\bar{G}}_{ee}(\bar{k}) = -i\omega\mu_0\bar{\bar{W}}_E^{-1} \tag{4.67a}$$

$$\bar{\bar{G}}_{em}(\bar{k}) = i\bar{\bar{W}}_E^{-1}\bar{\bar{k}}\bar{\bar{\mu}}^{-1} \tag{4.67b}$$

$$\bar{\bar{G}}_{me}(\bar{k}) = -i\bar{\bar{W}}_H^{-1}\bar{\bar{k}}\bar{\bar{\varepsilon}}^{-1} \tag{4.67c}$$

$$\bar{\bar{G}}_{mm}(\bar{k}) = -i\omega\varepsilon_0\bar{\bar{W}}_H^{-1} \tag{4.67d}$$

The relation between the inverses of the wave matrices is derived as

$$\bar{\bar{W}}_H^{-1} = \frac{1}{k_0^2}\left[\bar{\bar{\mu}}^{-1} - \bar{\bar{\mu}}^{-1}\bar{\bar{k}}\bar{\bar{W}}_E^{-1}\bar{\bar{k}}\bar{\bar{\mu}}^{-1}\right] \tag{4.68}$$

Equations (4.67)–(4.68) clearly show that the knowledge of the inverse of one type of wave matrix is sufficient to obtain the complete set of dyadic Green's functions in the k-domain. The final form of the dyadic Green's functions, which is valid everywhere but the source point, is obtained by substituting (4.67) into (4.50) as follows.

$$\bar{\bar{G}}(\bar{r},\bar{r}') = \frac{1}{(2\pi)^3}\int \begin{bmatrix} -i\omega\mu_0\bar{\bar{W}}_E^{-1} & i\bar{\bar{W}}_E^{-1}\bar{\bar{k}}\bar{\bar{\mu}}^{-1} \\ -i\bar{\bar{W}}_H^{-1}\bar{\bar{k}}\bar{\bar{\varepsilon}}^{-1} & -i\omega\varepsilon_0\bar{\bar{W}}_H^{-1} \end{bmatrix} e^{i\bar{k}.(\bar{r}-\bar{r}')}d^3\bar{k} \tag{4.69}$$

4.6 Dyadic Green's Functions for Uniaxially Anisotropic Medium

We apply the general form of the DGF obtained in Sect. 4.4, (4.69) to find DGF of an unbounded and layered uniaxially anisotropic medium. The problem of finding the complete set of DGFs for any type of anisotropic medium is simplified by (4.67)–(4.69) to finding the inverse of an electric wave matrix $\bar{\bar{W}}_E$ which is equal to

$$\bar{\bar{W}}_E^{-1} = \frac{adj\left(\bar{\bar{W}}_E\right)}{\left|\bar{\bar{W}}_E\right|} \tag{4.70}$$

$adj\left(\bar{\bar{W}}_E\right)$ is known as the adjoint and $\left|\bar{\bar{W}}_E\right|$ is known as the determinant of the electric wave matrix.

4.6.1 Dyadic Green's Functions for Unbounded Uniaxially Anisotropic Medium

In this section, the method outlined in Sect. 4.4, using the results given in Chap. 2, Sect. 2.2.1, is applied to obtain DGF for a uniaxially anisotropic medium.

The integral form of the electric type DGF for a uniaxially anisotropic medium is given by (4.69) and equals to

$$\bar{\bar{G}}_{ee}(\bar{r},\bar{r}') = \frac{-i\omega\mu_0}{(2\pi)^3}\int \bar{\bar{W}}_E^{-1} e^{i\bar{k}.(\bar{r}-\bar{r}')} d^3\bar{k} \tag{4.71}$$

When (4.70) is substituted into (4.71), we obtain

$$\bar{\bar{G}}_{ee}(\bar{r},\bar{r}') = \frac{-i\omega\mu_0}{(2\pi)^3}\int \frac{adj\left(\bar{\bar{W}}_E\right)}{\left|\bar{\bar{W}}_E\right|} e^{i\bar{k}.(\bar{r}-\bar{r}')} d^3\bar{k} \tag{4.72}$$

The dispersion relation for a uniaxially anisotropic medium from Chap. 2 was obtained as,

$$\left|\bar{\bar{W}}_E\right| = k_0^2\left(k^2 - k_0^2\varepsilon_{11}\right)\left[k_\rho^2\varepsilon_{11} + k_z^2\varepsilon_{33} - k_0^2\varepsilon_{11}\varepsilon_{33}\right] = 0 \tag{4.73a}$$

or

$$\left|\bar{\bar{W}}_E\right| = k_0^2\varepsilon_{33}\left(k_z^2 - \left(k_0^2\varepsilon_{11} - k_\rho^2\right)\right)\left[k_z^2 - \left(k_0^2\varepsilon_{11} - k_\rho^2\frac{\varepsilon_{11}}{\varepsilon_{33}}\right)\right] = 0 \tag{4.73b}$$

The wave numbers for a uniaxially anisotropic medium in Chap. 2 were obtained as

$$k_{zI}^2 = k_0^2\varepsilon_{11} - k_\rho^2 \tag{4.74a}$$

$$k_{zII}^2 = k_0^2\varepsilon_{11} - k_\rho^2\frac{\varepsilon_{11}}{\varepsilon_{33}} \tag{4.74b}$$

where

$$k^2 = k_\rho^2 + k_z^2 = k_x^2 + k_y^2 + k_z^2 \tag{4.75}$$

Equation (4.73b) can be put in the following form using (4.74)–(4.75)

$$\left|\bar{\bar{W}}_E\right| = k_0^2\varepsilon_{33}\left(k_z^2 - k_{zI}^2\right)\left(k_z^2 - k_{zII}^2\right) \tag{4.76}$$

$adj\left(\bar{\bar{W}}_E\right)$ for a uniaxially anisotropic medium is obtained in Chap. 2 and equal to

$$adj\left(\bar{\bar{W}}_E\right) = \left(k_0^2\varepsilon_{11} - k^2\right)\left[k_0^2\varepsilon_{11}\bar{\bar{I}} - \bar{k}\bar{k} - k_0^2(\varepsilon_{33} - \varepsilon_{11})\hat{p}\hat{p}\right] + k_0^2(\varepsilon_{33} - \varepsilon_{11}) \times (\bar{k} \times \hat{p})(\bar{k} \times \hat{p}) \tag{4.77}$$

$adj\left(\bar{\bar{W}}_E\right)$ can be put in the following matrix form

$$adj\left(\bar{\bar{W}}_E\right) = \begin{bmatrix} A_{11} & A_{12} & A_{13} \\ A_{21} & A_{22} & A_{23} \\ A_{31} & A_{32} & A_{33} \end{bmatrix} \tag{4.78}$$

The matrix elements of $adj\left(\bar{\bar{W}}_E\right)$ are derived as

$$A_{11} = k^2k_x^2 - k_0^2\left(k_x^2 + k_z^2\right)\varepsilon_{33} - k_\rho^2k_0^2\varepsilon_{11} + k_0^4\varepsilon_{11}\varepsilon_{33} \tag{4.79a}$$

$$A_{12} = A_{21} = k^2k_xk_y - k_0^2k_xk_y\varepsilon_{33} \tag{4.79b}$$

$$A_{13} = A_{31} = k^2k_xk_z - k_0^2k_xk_z\varepsilon_{11} \tag{4.79c}$$

$$A_{22} = k^2\left(k_y^2 - k_0^2\varepsilon_{33}\right) + k_0^2\left(k_x^2[\varepsilon_{33} - \varepsilon_{11}] - k_y^2\varepsilon_{11}\right) + k_0^4\varepsilon_{11}\varepsilon_{33} \tag{4.79d}$$

$$A_{23} = A_{32} = k^2k_yk_z - k_0^2k_yk_z\varepsilon_{11} \tag{4.79e}$$

$$A_{33} = k^2\left(k_z^2 - k_0^2\varepsilon_{11}\right) - k_0^2k_z^2\varepsilon_{11} + k_0^4\varepsilon_{11}^2 \tag{4.79f}$$

When (4.76)–(4.78) are substituted into (4.72), we obtain electric type DGF as

$$\bar{\bar{G}}_{ee}(\bar{r}, \bar{r}') = \frac{-i\omega\mu_0}{(2\pi)^3} \times \int \frac{1}{k_0^2\varepsilon_{33}\left(k_z^2 - k_{zI}^2\right)\left(k_z^2 - k_{zII}^2\right)} \begin{bmatrix} A_{11} & A_{12} & A_{13} \\ A_{21} & A_{22} & A_{23} \\ A_{31} & A_{32} & A_{33} \end{bmatrix} e^{i\bar{k}.(\bar{r}-\bar{r}')}dk_xdk_ydk_z \tag{4.80}$$

We perform the integration over k_z to obtain the two dimensional form of the DGF that is widely used in radiation and scattering problems. This form of the DGF is useful especially for the layered structures when the stratification of the layers is in the z-direction. The poles of the integrand occur at the zeros of $|\bar{\bar{W}}_E|$ denoted by $k_z = \pm k_{zI}$ and $k_z = \pm k_{zII}$. We assume the medium to be slightly lossy, i.e. $\mathrm{Im}k_z << \mathrm{Re}k_z$, $\mathrm{Im}k_z > 0$. This guarantees that the radiation condition is satisfied at $z = \pm\infty$. When we perform the contour integration over k_z, we obtain the following result for z > z′

$$\bar{\bar{G}}_{ee}(\bar{r},\bar{r}') = \frac{\omega\mu_0}{4\pi^2 k_0{}^2} \int_{-\infty}^{\infty}\int_{-\infty}^{\infty} dk_x dk_y \left\{ \frac{adj\bar{\bar{W}}_{EI}}{2k_{zI}(k_{zI}^2 - k_{zII}^2)\varepsilon_{33}} e^{i\bar{k}_I\cdot(\bar{r}-\bar{r}')} + \frac{adj\bar{\bar{W}}_{EII}}{2k_{zII}(k_{zII}^2 - k_{zI}^2)\varepsilon_{33}} e^{i\bar{k}_{II}\cdot(\bar{r}-\bar{r}')} \right]\Bigg\}, \quad z>z' \tag{4.81}$$

In (4.81), $adj\bar{\bar{W}}_{EI}$ in matrix form can be obtained using (4.78)–(4.79) as

$$adj\bar{\bar{W}}_{EI} = adj\bar{\bar{W}}_E\Big|_{k_z=k_{zI}} \tag{4.82}$$

So,

$$adj\bar{\bar{W}}_{EI} = \begin{bmatrix} B_{11} & B_{12} & B_{13} \\ B_{21} & B_{22} & B_{23} \\ B_{31} & B_{32} & B_{33} \end{bmatrix} \tag{4.83}$$

The matrix elements are

$$B_{11} = k_0^2 k_y^2(\varepsilon_{33} - \varepsilon_{11}) \tag{4.84a}$$

$$B_{12} = B_{21} = k_0^2 k_x k_y(\varepsilon_{11} - \varepsilon_{33}) \tag{4.84b}$$

$$B_{13} = B_{31} = 0 \tag{4.84c}$$

$$B_{22} = k_0^2 k_x^2(\varepsilon_{33} - \varepsilon_{11}) \tag{4.84d}$$

$$B_{23} = B_{32} = 0 \tag{4.84e}$$

$$B_{33} = 0 \tag{4.84f}$$

Following the same procedure, $adj\bar{\bar{W}}_{EII}$ in matrix form can be obtained as

$$adj\bar{\bar{W}}_{EII} = adj\bar{\bar{W}}_E\Big|_{k_z=k_{zII}} \tag{4.85}$$

So,

$$adj\bar{\bar{W}}_{EII} = \begin{bmatrix} C_{11} & C_{12} & C_{13} \\ C_{21} & C_{22} & C_{23} \\ C_{31} & C_{32} & C_{33} \end{bmatrix} \tag{4.86}$$

The matrix elements for $adj\bar{\bar{W}}_{EII}$ are

$$C_{11} = k_x^2 k_{zII}^2 \left(\frac{\varepsilon_{11} - \varepsilon_{33}}{\varepsilon_{11}}\right) \tag{4.87a}$$

$$C_{12} = C_{21} = k_x k_y k_{zII}^2 \left(\frac{\varepsilon_{11} - \varepsilon_{33}}{\varepsilon_{11}}\right) \tag{4.87b}$$

$$C_{13} = C_{31} = k_x k_{zII} \left(\frac{\varepsilon_{11} - \varepsilon_{33}}{\varepsilon_{11}}\right)\left(-k_\rho^2 \frac{\varepsilon_{11}}{\varepsilon_{33}}\right) \tag{4.87c}$$

$$C_{22} = k_y^2 k_{zII}^2 \left(\frac{\varepsilon_{11} - \varepsilon_{33}}{\varepsilon_{11}}\right) \tag{4.87d}$$

$$C_{23} = C_{32} = k_y k_{zII} \left(\frac{\varepsilon_{11} - \varepsilon_{33}}{\varepsilon_{11}}\right)\left(-k_\rho^2 \frac{\varepsilon_{11}}{\varepsilon_{33}}\right) \tag{4.87e}$$

$$C_{33} = \left(\frac{\varepsilon_{11} - \varepsilon_{33}}{\varepsilon_{11}}\right)\left(k_\rho^2 \frac{\varepsilon_{11}}{\varepsilon_{33}}\right)^2 \tag{4.87f}$$

The final form of the DGF given in (4.61) check with the results given in [1] when the optic axis is in the z-direction, i.e., tilt angle is zero $\psi = 0°$. $\bar{\bar{G}}(\bar{r},\bar{r}')$ for $z<z'$ can be obtained assuming $\mathrm{Im}(-k_z^{\circ})<0$ and $\mathrm{Im}(k_z^{ed})<0$. The DGF given in (4.81) can be written in terms of ordinary and extraordinary waves as

$$\begin{aligned}\bar{\bar{G}}(\bar{r},\bar{r}') =& \frac{i}{8\pi^2}\iint_{-\infty}^{\infty} dk_x dk_y \left\{\frac{1}{k_z^{\circ}}\hat{o}(k_z^{\circ})\hat{o}(k_z^{\circ})e^{i\bar{k}^{\circ}\cdot(\bar{r}-\bar{r}')}\right. \\ &\left. + \frac{2}{k_z^{eu} - k_z^{ed}} \cdot \frac{\omega^2\mu(\varepsilon+\varepsilon_z) - (k_x^2 + k_y^2 + k_z^{eu2})}{\omega^2\mu\varepsilon_{33}} \hat{e}(k_z^{eu})\hat{e}(k_z^{eu})e^{i\bar{k}^{e}\cdot(\bar{r}-\bar{r}')}\right\}, \quad z>z'\end{aligned} \tag{4.88a}$$

and

$$\begin{aligned}\bar{\bar{G}}(\bar{r},\bar{r}') =& \frac{i}{8\pi^2}\iint_{-\infty}^{\infty} dk_x dk_y \left\{\frac{1}{k_z^{\circ}}\hat{o}(-k_z^{\circ})\hat{o}(-k_z^{\circ})e^{i\bar{\kappa}^{\circ}\cdot(\bar{r}-\bar{r}')}\right. \\ &\left. + \frac{2}{k_z^{eu} - k_z^{ed}} \cdot \frac{\omega^2\mu(\varepsilon+\varepsilon_z) - (k_x^2 + k_y^2 + k_z^{eu2})}{\omega^2\mu\varepsilon_{33}} \hat{e}(k_z^{ed})\hat{e}(k_z^{ed})e^{i\bar{\kappa}^{e}\cdot(\bar{r}-\bar{r}')}\right\}, \quad z<z'\end{aligned} \tag{4.88b}$$

where $\widehat{o}(\pm k_z^{\circ})$ is a unit vector for the an ordinary wave and $\widehat{e}(k_z^{eu})$ or $\widehat{e}(k_z^{ed})$ is the unit vector for an extraordinary wave and they are defined in the Appendix of Chap. 5. The vectors are defined as

$$\bar{\kappa}^{\circ} = \bar{k}_{\rho} - k_z^{\circ}\widehat{z} \tag{4.89a}$$

$$\bar{\kappa}^{e} = \bar{k}_{\rho} + k_z^{ed}\widehat{z} \tag{4.89b}$$

$$\bar{k}_u = \bar{\bar{\varepsilon}} \cdot \bar{k}^e \tag{4.89c}$$

$$\bar{\kappa}_u = \bar{\bar{\varepsilon}} \cdot \bar{\kappa}^e \tag{4.89d}$$

$$\bar{k}_{\rho} = k_x\widehat{x} + k_y\widehat{y} \tag{4.89e}$$

4.6.2 *Dyadic Green's Functions for Layered Uniaxially Anisotropic Medium*

Dyadic Green's function for a multilayered anisotropic medium can be obtained by considering a source point located in Region 0 above a stratified medium that is illustrated in Fig. 4.2 below.

The medium in Region l is assumed to be uniaxially anisotropic which is characterized by permittivity tensor $\bar{\bar{\varepsilon}}_l$, $l = 1,2,\ldots,n$. The medium at the top and bottom of the stratification are assumed to be isotropic with permittivities ε_0 and ε_t. The permeability μ is common to all media. The first subscript l is used to denote the region of the observation point $\bar{r}$ and the second subscript j is used to denote the region of the source point $\bar{r}'$ for DGF $\bar{\bar{G}}_{lj}$. The DGFs in each layer can be formulated as follows [3,4].

$$\begin{aligned}\bar{\bar{G}}_{00}(\bar{r},\bar{r}') = \frac{i}{8\pi^2}\iint dk_x dk_y \frac{1}{k_{0z}}\Big\{&\Big[\widehat{h}_0(-k_{0z})e^{i(\bar{\kappa}_0\cdot\bar{r})} + R_{HH}\widehat{h}_0(k_{0z})e^{i(\bar{k}_0\cdot\bar{r})}\\ &+ R_{HV}\widehat{v}_0(k_{0z})e^{i(\bar{k}_0\cdot\bar{r})}\Big]\widehat{h}_0(-k_{0z}) + \Big[\widehat{v}_0(-k_{0z})e^{i(\bar{\kappa}_0\cdot\bar{r})} + R_{VV}\widehat{v}_0(k_{0z})e^{i(\bar{k}_0\cdot\bar{r})}\\ &+ R_{VH}\widehat{h}_0(k_{0z})e^{i(\bar{k}_0\cdot\bar{r})}\Big]\widehat{v}_0(-k_{0z})\Big\}e^{-i(\bar{\kappa}_0\cdot\bar{r}')}\end{aligned} \tag{4.90a}$$

$$\begin{aligned}\bar{\bar{G}}_{l0}(\bar{r},\bar{r}') = \frac{i}{8\pi^2}\iint dk_x dk_y \frac{1}{k_{0z}}\Big\{&\big[A_{lHo}\widehat{o}(-k_{1z}^{o})e^{i(\bar{\kappa}_l^{o}\cdot\bar{r})}\\ &+ B_{lHo}\widehat{o}(k_{1z}^{o})e^{i(\bar{k}_l^{o}\cdot\bar{r})} + B_{lHe}\widehat{e}(k_{1z}^{eu})e^{i(\bar{k}_l^{e}\cdot\bar{r})} + A_{lHe}\widehat{e}(k_{1z}^{ed})e^{i(\bar{\kappa}_l^{e}\cdot\bar{r})}\big]\widehat{h}_0(-k_{0z})\\ &+ \big[A_{lVo}\widehat{o}(-k_{1z}^{o})e^{i(\bar{\kappa}_l^{o}\cdot\bar{r})} + B_{lVo}\widehat{o}(k_{1z}^{o})e^{i(\bar{k}_l^{o}\cdot\bar{r})} + B_{lVe}\widehat{e}(k_{1z}^{eu})e^{i(\bar{k}_l^{e}\cdot\bar{r})}\\ &+ A_{lVe}\widehat{e}(k_{1z}^{ed})e^{i(\bar{\kappa}_l^{e}\cdot\bar{r})}\big]\widehat{v}_0(-k_{0z})\Big\}e^{-i(\bar{\kappa}_0\cdot\bar{r}')} \quad l = 1,2,\ldots,n\end{aligned} \tag{4.90b}$$

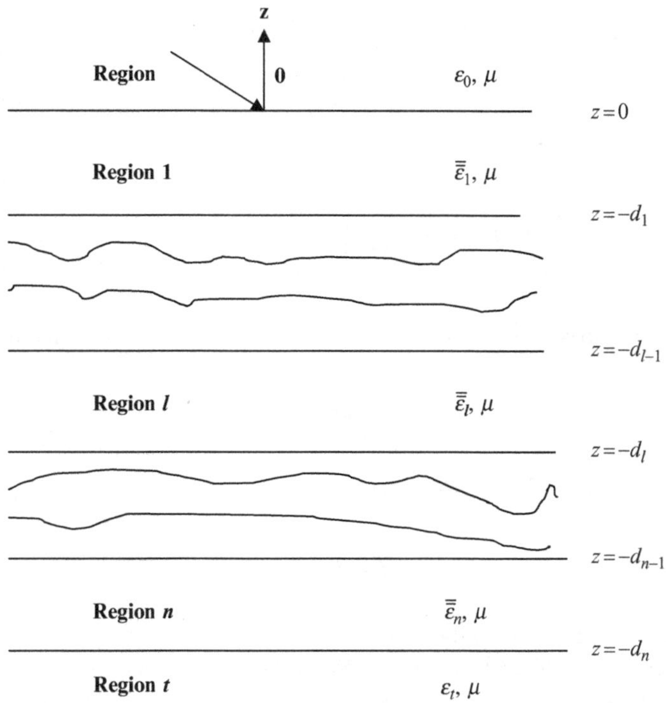

Fig. 4.2 Geometric configuration of multi-layered anisotropic media

$$\bar{\bar{G}}_{t0}(\bar{r},\bar{r}') = \frac{i}{8\pi^2}\iint dk_x dk_y \frac{1}{k_{0z}}\left\{\left[X_{HH}\hat{h}_t(-k_{tz}) + X_{HV}\hat{v}_t(-k_{tz})\right]\hat{h}_o(-k_{0z})\right.$$
$$\left. + \left[X_{VV}\hat{v}_t(-k_{tz}) + X_{VH}\hat{h}_t(-k_{tz})\right]\hat{v}_o(-k_{0z})\right\} e^{i(\bar{\kappa}_t\cdot\bar{r})} e^{-i(\bar{\kappa}_o\cdot\bar{r}')} \quad (4.90c)$$

where $\hat{h}$ and $\hat{v}$ represent the horizontally and vertically polarized waves, respectively. The particular form in (4.90) is based on the result of (4.88). The coefficients, $R_{HH}, R_{HV}, R_{VH} R_{VV}, A_{lHo}, B_{lHo}, A_{lHe}, B_{lHe}, A_{lVo}, B_{lVo}, A_{lVe}, B_{lVe}, X_{HH}, X_{HV}, X_{VH}$ and X_{VV} are related through boundary conditions.

4.7 Dyadic Green's Functions for Gyrotropic Medium

In this section, the complete set of DGFs for unbounded electrically gyrotropic or gyroelectric medium will be obtained following the steps outlined in Sect. 4.4.

4.7.1 Electric Type DGF $\bar{\bar{G}}_{ee}(\bar{r},\bar{r}')$ for a Gyroelectric Medium

Electric type DGF, $\bar{\bar{G}}_{ee}(\bar{r},\bar{r}')$ is the solution of the second order dyadic differential equation given by (4.29). The integral form of the electric type DGF for a gyroelectric medium can be written using (4.69) as

$$\bar{\bar{G}}_{ee}(\bar{r},\bar{r}') = \frac{-i\omega\mu_0}{(2\pi)^3}\int \bar{\bar{W}}_E^{-1} e^{i\bar{k}.(\bar{r}-\bar{r}')} d^3\bar{k} \tag{4.91}$$

or

$$\bar{\bar{G}}_{ee}(\bar{r},\bar{r}') = \frac{-i\omega\mu_0}{(2\pi)^3}\int \frac{adj\left(\bar{\bar{W}}_E\right)}{\left|\bar{\bar{W}}_E\right|} e^{i\bar{k}.(\bar{r}-\bar{r}')} d^3\bar{k} \tag{4.92}$$

Electric wave matrix of gyroelectric medium in Chap. 2 was obtained as

$$\bar{\bar{W}}_E = \left[\bar{\bar{k}}\bar{\bar{k}} + k_0{}^2\bar{\bar{\varepsilon}}\right] \tag{4.93a}$$

The dispersion relation for gyroelectric medium is

$$|\bar{\bar{W}}_E| = k_0{}^2\varepsilon_3(k_z{}^2 - k_{zI}{}^2)(k_z{}^2 - k_{zII}{}^2) = 0 \tag{4.93b}$$

To facilitate the derivation of the adjoint of an electric wave matrix, we will operate with dyadics as described in Chap. 1. When the relative permittivity tensor for a gyroelectric medium is substituted into (4.93a), we obtain the electric wave matrix in dyadic form as follows.

$$\bar{\bar{W}}_E = (k_0^2\varepsilon_1 - k^2)\bar{\bar{I}} + \hat{b}_0\hat{b}_0\left[k_0{}^2(\varepsilon_3 - \varepsilon_1)\right] + ik_0{}^2\varepsilon_2(\hat{b}_0 \times \bar{\bar{I}}) + \bar{k}\bar{k} \tag{4.94}$$

Using the dyadic identity,

$$\begin{aligned} adj\left[\lambda\bar{\bar{I}} + \hat{c}\times\bar{\bar{I}} + \hat{a}\hat{l} + \hat{b}\hat{m}\right] = {} & [\lambda^2 + (\hat{a}\cdot\hat{l} + \hat{b}\cdot\hat{m})\lambda + \hat{b}\times\hat{c}\cdot\hat{m} + \hat{a}\times\hat{c}\cdot\hat{l}]\bar{\bar{I}} \\ & - (\lambda + \hat{a}\cdot\hat{l} + \hat{b}\cdot\hat{m})(\hat{c}\times\bar{\bar{I}}) - \lambda(\hat{a}\hat{l} + \hat{b}\hat{m}) \\ & + (\hat{l}\times\hat{m})(\hat{a}\times\hat{b}) + \hat{c}\hat{c} + (\hat{c}\times\hat{a})\hat{l} + (\hat{c}\times\hat{b})\hat{m} \\ & + \hat{a}(\hat{l}\times\hat{c}) + \hat{b}(\hat{m}\times\hat{c}) \end{aligned} \tag{4.95}$$

The adjoint of $\bar{\bar{W}}_E$ or $adj\left(\bar{\bar{W}}_E\right)$ can be written as

$$\begin{aligned} adj\bar{\bar{W}}_E = {} & (k_0{}^4 adj\bar{\bar{\varepsilon}} - k^2k_0{}^2\varepsilon_3\bar{\bar{I}}) + \hat{k}\hat{k}\left[k^2(k^2 - k_0{}^2\varepsilon_1)\right] + \hat{b}_0\hat{b}_0\left[k^2k_0{}^2(\varepsilon_3 - \varepsilon_1)\right] \\ & + (\hat{k}\times\hat{b}_0)(\hat{k}\times\hat{b}_0)[k^2k_0{}^2(\varepsilon_3 - \varepsilon_1)] + i\varepsilon_2k^2k_0{}^2\left[\hat{k}(\hat{k}\times\hat{b}_0) - (\hat{k}\times\hat{b}_0)\hat{k}\right] \end{aligned} \tag{4.96}$$

We can represent $adj\left(\bar{\bar{W}}_E\right)$ in matrix form as,

$$adj\bar{\bar{W}}_E = \begin{bmatrix} A_{11} & A_{12} & A_{13} \\ A_{21} & A_{22} & A_{23} \\ A_{31} & A_{32} & A_{33} \end{bmatrix} \tag{4.97}$$

The elements of the matrix in (4.97) are

$$A_{11} = (k_\rho{}^2 + k_z{}^2)k_x{}^2 - k_0{}^2[\varepsilon_1 k_\rho{}^2 + \varepsilon_3(k_x{}^2 + k_z{}^2)] + k_0{}^4\varepsilon_1\varepsilon_3 \tag{4.98a}$$

$$A_{12} = (k_\rho{}^2 + k_z{}^2)k_x k_y - k_0{}^2[i\varepsilon_2 k_\rho{}^2 + \varepsilon_3 k_x k_y] + ik_0{}^4\varepsilon_2\varepsilon_3 \tag{4.98b}$$

$$A_{13} = (k_\rho{}^2 + k_z{}^2)k_x k_z - k_0{}^2[\varepsilon_1 k_x k_z + i\varepsilon_2 k_y k_z] \tag{4.98c}$$

$$A_{21} = (k_\rho{}^2 + k_z{}^2)k_x k_y - k_0{}^2[-i\varepsilon_2 k_\rho{}^2 + \varepsilon_3 k_x k_y] - ik_0{}^4\varepsilon_2\varepsilon_3 \tag{4.98d}$$

$$A_{22} = (k_\rho{}^2 + k_z{}^2)k_y{}^2 - k_0{}^2[\varepsilon_1 k_\rho{}^2 + \varepsilon_3(k_y{}^2 + k_z{}^2)] + k_0{}^4\varepsilon_1\varepsilon_3 \tag{4.98e}$$

$$A_{23} = (k_\rho{}^2 + k_z{}^2)k_y k_z - k_0{}^2[\varepsilon_1 k_y k_z - i\varepsilon_2 k_x k_z] \tag{4.98f}$$

$$A_{31} = (k_\rho{}^2 + k_z{}^2)k_x k_z - k_0{}^2[\varepsilon_1 k_x k_z - i\varepsilon_2 k_y k_z] \tag{4.98g}$$

$$A_{32} = (k_\rho{}^2 + k_z{}^2)k_y k_z - k_0{}^2[\varepsilon_1 k_y k_z + i\varepsilon_2 k_x k_z] \tag{4.98h}$$

$$A_{33} = (k_\rho{}^2 + k_z{}^2)k_z{}^2 - k_0{}^2[\varepsilon_1(k_\rho{}^2 + 2k_z{}^2)] + k_0{}^4[\varepsilon_1{}^2 - \varepsilon_2{}^2] \tag{4.98i}$$

where

$$k_\rho{}^2 = k_x{}^2 + k_y{}^2 \tag{4.99}$$

We perform the integration over k_z after substituting (4.73), (4.97) into (4.92). The poles of the integrand occur at the zeros of $|\bar{\bar{W}}_E|$ denoted by $k_z = \pm k_{zI}$ and $k_z = \pm k_{zII}$ where k_{zI} and k_{zII} are defined by (3.62) and (3.63), respectively. Assuming the medium to be slightly lossy, i.e. $\mathrm{Im}k_z << \mathrm{Re}k_z, \mathrm{Im}k_z > 0$ and performing the contour integration over k_z, we obtain the following result for z > z′:

$$\bar{\bar{G}}_{ee}(\bar{r},\bar{r}') = \frac{-\omega\mu_0}{8\pi^2}\int_{-\infty}^{\infty}\int_{-\infty}^{\infty} dk_x dk_y \left\{ \frac{1}{k_0{}^2\varepsilon_3(k_{zI}{}^2 - k_{zII}{}^2)} \left[\frac{adj\bar{\bar{W}}_E(k_{zI})}{k_{zI}} e^{i\bar{\kappa}_I\cdot(\bar{r}-\bar{r}')} - \frac{adj\bar{\bar{W}}_E(k_{zII})}{k_{zII}} e^{i\bar{\kappa}_{II}\cdot(\bar{r}-\bar{r}')} \right] \right\}, \quad z < z' \tag{4.100}$$

Similarly, when z < z′, $\bar{\bar{G}}_{ee}(\bar{r},\bar{r}')$ can be obtained by assuming $\mathrm{Im}(-k_{zI})<0$ and $\mathrm{Im}(-k_{zII})<0$ as

$$\bar{\bar{G}}_{ee}(\bar{r},\bar{r}')=\frac{-\omega\mu_0}{8\pi^2}\int_{-\infty}^{\infty}\int_{-\infty}^{\infty}dk_xdk_y\left\{\frac{1}{k_0{}^2\varepsilon_3(k_{zI}{}^2-k_{zII}{}^2)}\left[\frac{adj\bar{\bar{W}}_E(-k_{zI})}{k_{zI}}e^{i\bar{\kappa}_I\cdot(\bar{r}-\bar{r}')}-\frac{adj\bar{\bar{W}}_E(k_{zII})}{k_{zII}}e^{i\bar{\kappa}_{II}\cdot(\bar{r}-\bar{r}')}\right]\right\},\ z<z' \tag{4.101}$$

where

$$\bar{k}_I=\bar{k}_\rho+\hat{z}k_{zI} \tag{4.102a}$$

$$\bar{k}_{II}=\bar{k}_\rho+\hat{z}k_{zII} \tag{4.102b}$$

$$\bar{\kappa}_I=\bar{k}_\rho-\hat{z}k_{zI} \tag{4.102c}$$

$$\bar{\kappa}_{II}=\bar{k}_\rho-\hat{z}k_{zII} \tag{4.102d}$$

$\bar{k}_I$, $\bar{k}_{II}$ represent the wave vectors for the upward (+z) traveling waves of type I and type II. $\bar{\kappa}_I$, $\bar{\kappa}_{II}$ represent those for the downward (−z) traveling waves.

We can represent the DGFs given by (4.101) and (4.102) in dyadic form by finding the eigenvalues and eigenvectors of the adjoint matrix for $\bar{\bar{W}}_E$, i.e., $adj\left(\bar{\bar{W}}_E\right)$. If we review the elements of the adjoint matrix $adj\left(\bar{\bar{W}}_E\right)$ given by (4.98), we see that

$$adj\left(\bar{\bar{W}}_E\right)=\left(adj\left(\bar{\bar{W}}_E\right)\right)^{\dagger}. \tag{4.103}$$

where † denotes the conjugate transpose of the matrix. This requires $adj\left(\bar{\bar{W}}_E\right)$ to be a *Hermitian* matrix. The eigenvalues of a Hermitian matrix are real and the eigenvectors corresponding to distinct eigenvalues are orthogonal in the sense that the Hermitian dot product vanishes. In other words every Hermitian matrix possesses a complete set of orhonormal eigenvectors. In this case, the completeness relation [2] becomes

$$\bar{\bar{I}}=\hat{u}_1\hat{u}_1^*+\hat{u}_2\hat{u}_2^*+\hat{u}_3\hat{u}_3^* \tag{4.104}$$

where

$$\hat{u}_i^*\bullet\hat{u}_j=\delta_{ij} \tag{4.105}$$

and $\hat{u}_1$, $\hat{u}_2$ and $\hat{u}_3$ are the orthonormal eigenvectors of the Hermitian matrix. Then the dyadic decomposition of the matrix $adj\left(\bar{\bar{W}}_E\right)$ takes the form as

$$adj\left(\bar{\bar{W}}_E\right) = \lambda_1\hat{u}_1\hat{u}_1^* + \lambda_2\hat{u}_2\hat{u}_2^* + \lambda_3\hat{u}_3\hat{u}_3^* \tag{4.106}$$

Hence at this point we proved that we can write the DGFs given by (4.101)–(4.102) in dyadic form. To simplify our analysis and find the eigenvalues and the eigenvectors of the matrix $adj\left(\bar{\bar{W}}_E\right)$, we will investigate the characteristic equation of $adj\left(\bar{\bar{W}}_E\right)$. The characteristic equation of $adj\left(\bar{\bar{W}}_E\right)$, i.e., $f(\lambda)$ can be expressed as

$$f(\lambda) = \left|\lambda I - adj\bar{\bar{W}}_E\right| = \lambda^3 - tr(adj\bar{\bar{W}}_E)\lambda^2 + tr(adj(adj\bar{\bar{W}}_E)\lambda - \left|adj\bar{\bar{W}}_E\right| = 0 \tag{4.107}$$

where tr stands for the trace of the matrix. Using the following identities,

$$\left|adj\bar{\bar{W}}_E\right| = \left|\bar{\bar{W}}_E\right|^2 \tag{4.108}$$

$$adj(adj\bar{\bar{W}}_E) = \left|\bar{\bar{W}}_E\right|\bar{\bar{W}}_E \tag{4.109}$$

$f(\lambda)$ can be rewritten as

$$f(\lambda) = \lambda^3 - tr\left(adj\bar{\bar{W}}_E\right)\lambda^2 + tr\left(\left|\bar{\bar{W}}_E\right|\bar{\bar{W}}_E\right)\lambda - \left|\bar{\bar{W}}_E\right|^2 = 0 \tag{4.110}$$

Since $|\bar{\bar{W}}_E(k_z)|$ is zero when $k_z = \pm k_{zI}$ or $k_z = \pm k_{zII}$, then characteristic equation for $adj(\bar{\bar{W}}_E)$ reduces to

$$f(\lambda) = \lambda^3 - tr\left(adj\bar{\bar{W}}_E\right)\lambda^2 \tag{4.111}$$

Hence, the eigenvalues for $adj\left(\bar{\bar{W}}_E\right)$ are

$$\lambda_1 = tr\left(adj\bar{\bar{W}}_E\right), \quad \lambda_2 = \lambda_3 = 0 \tag{4.112}$$

As a result, using (4.106) we can express $adj\bar{\bar{W}}_E$ as a single dyad in the following form for the adjoint matrices of the type I and the type II waves as follows.

$$adj\bar{\bar{W}}_E(\pm k_{zI}) = \alpha_I\left[\hat{e}_{nI}(\pm k_{zI})\hat{e}_{nI}^*(\pm k_{zI})\right] \tag{4.113a}$$

$$adj\bar{\bar{W}}_E(\pm k_{zII}) = \alpha_{II}\left[\hat{e}_{nII}(\pm k_{zII})\hat{e}_{nII}^*(\pm k_{zII})\right] \tag{4.113b}$$

$\lambda_I = \alpha_I$, $\lambda_{II} = \alpha_{II}$ are the eigenvalues and are defined as

$$\alpha_I = k_I^4 - k_I^2 {k_0}^2 \left[\varepsilon_1 \left(3 - \cos^2\theta\right) + \varepsilon_3 \left(1 + \cos^2\theta\right)\right] + {k_0}^4 \left({\varepsilon_1}^2 - {\varepsilon_2}^2 + 2\varepsilon_1\varepsilon_3\right) \tag{4.114a}$$

$$\alpha_{II} = k_{II}^4 - k_{II}^2 {k_0}^2 \left[\varepsilon_1 \left(3 - \cos^2\theta\right) + \varepsilon_3 \left(1 + \cos^2\theta\right)\right] + {k_0}^4 \left({\varepsilon_1}^2 - {\varepsilon_2}^2 + 2\varepsilon_1\varepsilon_3\right) \tag{4.114b}$$

$\hat{e}_{nI}(\pm k_{zI})$ and $\hat{e}_{nII}(\pm k_{zI})$ are the orthonormal eigenvectors representing two characteristic electric fields for the type I and type II waves that exist in a gyro-electric medium and they are defined as

$$\hat{e}_{nI}(\pm k_{zI}) = \frac{\bar{e}_I(\pm k_{zI})}{norm(\bar{e}_I(\pm k_{zI}))} \tag{4.115}$$

$$\hat{e}_{nII}(\pm k_{zII}) = \frac{\bar{e}_{II}(\pm k_{zII})}{norm(\bar{e}_{II}(\pm k_{zII}))} \tag{4.116}$$

where

$$\bar{e}_I(\pm k_{zI}) = \begin{bmatrix} 1 \\ \dfrac{A_{13}A_{21} + A_{23}\alpha_I - A_{23}A_{11}}{\alpha_I A_{13} - A_{22}A_{13} + A_{23}A_{12}} \\ \dfrac{-A_{12}}{A_{13}} \left[\dfrac{A_{13}A_{21} + A_{23}\alpha_I - A_{23}A_{11}}{\alpha_I A_{13} - A_{22}A_{13} + A_{13}A_{12}}\right] + \dfrac{\alpha_I - A_{11}}{A_{13}} \end{bmatrix} \tag{4.117}$$

$$\bar{e}_{II}(\pm k_{zI}) = \begin{bmatrix} 1 \\ \dfrac{A_{13}A_{21} + A_{23}\alpha_{II} - A_{23}A_{11}}{\alpha_{II} A_{13} - A_{22}A_{13} + A_{23}A_{12}} \\ \dfrac{-A_{12}}{A_{13}} \left[\dfrac{A_{13}A_{21} + A_{23}\alpha_{II} - A_{23}A_{11}}{\alpha_{II} A_{13} - A_{22}A_{13} + A_{13}A_{12}}\right] + \dfrac{\alpha_{II} - A_{11}}{A_{13}} \end{bmatrix} \tag{4.118}$$

where the elements of $\bar{e}_I(\pm k_{zI})$ and $\bar{e}_{II}(\pm k_{zII})$, $A_{ij}, (i,j) = 1, 2, 3$ are defined by (4.98). The form of $\bar{e}_I$, $\bar{e}_{II}$ given by (4.117), (4.118) are valid when the x component of the electric field is not zero. The eigenvectors given by (4.117) and (4.118) are found by solving the eigenvalue problem $adj\bar{\bar{W}}_E.\hat{u} = \lambda\hat{u}$. For each eigenvector, the corresponding eigenvalues are given by (4.112).

Now, we can represent the DGF $\bar{\bar{G}}_{ee}(\bar{r},\bar{r}')$ given by (4.100) in dyadic form when $z > z'$ as

$$\bar{\bar{G}}_{ee}(\bar{r},\bar{r}') = \frac{-\omega\mu_0}{8\pi^2} \underset{z<z'}{\int_{-\infty}^{\infty}\int_{-\infty}^{\infty}} dk_x dk_y \left\{ \frac{1}{k_0{}^2 \varepsilon_3 (k_{zI}{}^2 - k_{zII}{}^2)} \left[\frac{\alpha_I}{k_{zI}} \hat{e}_{nI}(k_{zI}) \hat{e}^*_{nI}(k_{zI}) e^{i\bar{\kappa}_I \cdot (\bar{r}-\bar{r}')} \right. \right. \\ \left. \left. \frac{\alpha_{II}}{k_{zII}} \hat{e}_{nII}(k_{zII}) \hat{e}^*_{nII}(k_{zII}) e^{i\bar{\kappa}_I \cdot (\bar{r}-\bar{r}')} \right] \right\}, \tag{4.119}$$

Similarly, the DGF $\bar{\bar{G}}_{ee}(\bar{r},\bar{r}')$ given by (4.101) is written in dyadic form when $z < z'$ as

$$\bar{\bar{G}}_{ee}(\bar{r},\bar{r}') = \frac{-\omega\mu_0}{8\pi^2} \underset{z<z'}{\int_{-\infty}^{\infty}\int_{-\infty}^{\infty}} dk_x dk_y \left\{ \frac{1}{k_0{}^2 \varepsilon_3 (k_{zI}{}^2 - k_{zII}{}^2)} \left[\frac{\alpha_I}{k_{zI}} \hat{e}_{nI}(-k_{zI}) \hat{e}^*_{nI}(-k_{zI}) e^{i\bar{\kappa}_I \cdot (\bar{r}-\bar{r}')} \right. \right. \\ \left. \left. - \frac{\alpha_{II}}{k_{zII}} \hat{e}_{nII}(k_{zII}) \hat{e}^*_{nII}(k_{zII}) e^{i\bar{\kappa}_I \cdot (\bar{r}-\bar{r}')} \right] \right\}, \tag{4.120}$$

The form of the DGFs given by (4.119) and (4.120) is specifically useful for the problems which have planar stratification. It also gives a good physical interpretation by exposing the polarization of the waves. This type of feature is useful especially for the radiation problems. It simplifies the analytical calculation by using only one vector to operate on the source and the other vector to define the polarization of the radiation field.

4.7.2 *Magnetic Type DGF $\bar{\bar{G}}_{mm}(\bar{r},\bar{r}')$ for a Gyroelectric Medium*

Magnetic type DGF, $\bar{\bar{G}}_{mm}(\bar{r},\bar{r}')$, is the solution of the second order dyadic differential equation given by (4.30). Following the same steps, we can write

$$\bar{\bar{G}}_{mm}(\bar{r},\bar{r}') = \frac{-i\omega\varepsilon_0}{(2\pi)^3} \int_{-\infty}^{\infty} \bar{\bar{W}}_H^{-1} e^{i\bar{k}.(\bar{r}-\bar{r}')} d^3\bar{k} \tag{4.121}$$

or

$$\bar{\bar{G}}^e_{mm}(\bar{r},\bar{r}') = \frac{-i\omega\varepsilon_0}{(2\pi)^3} \int_{-\infty}^{\infty} \frac{adj\left(\bar{\bar{W}}_H\right)}{\left|\bar{\bar{W}}_H\right|} e^{i\bar{k}.(\bar{r}-\bar{r}')} d^3\bar{k} \tag{4.122}$$

where

$$\bar{\bar{W}}_H^{-1} = \frac{adj\left(\bar{\bar{W}}_H\right)}{\left|\bar{\bar{W}}_H\right|} \tag{4.123}$$

Since

$$k_0^2 \bar{\bar{W}}_H^{-1} = \bar{\bar{I}} - \bar{\bar{k}} \bar{\bar{W}}_E^{-1} \bar{\bar{k}} \tag{4.124}$$

we can express $\bar{\bar{W}}_H^{-1}$ as

$$\bar{\bar{W}}_H^{-1} = \frac{1}{k_0^2} \frac{\left[\left|\bar{\bar{W}}_E\right| \bar{\bar{I}} - \bar{\bar{k}} adj \bar{\bar{W}}_E \bar{\bar{k}}\right]}{\left|\bar{\bar{W}}_E\right|} \tag{4.125}$$

We can now rewrite $\bar{\bar{W}}_H^{-1}$ in matrix form as

$$\bar{\bar{W}}_H^{-1} = \frac{\bar{\bar{B}}}{\left|\bar{\bar{W}}_E\right|} \tag{4.126}$$

where

$$\bar{\bar{B}} = \begin{bmatrix} B_{11} & B_{12} & B_{13} \\ B_{21} & B_{22} & B_{23} \\ B_{31} & B_{32} & B_{33} \end{bmatrix} \tag{4.127}$$

The elements of the matrix $\bar{\bar{B}}$ in (4.127) are given by

$$\begin{aligned}
B_{11} &= k_z^2 k_x^2 \varepsilon_3 + k_0^4 \varepsilon_3 (\varepsilon_1^2 - \varepsilon_2^2) - k_0^2 \left[k_x^2 (\varepsilon_1^2 - \varepsilon_2^2) + (k_\rho^2 + k_z^2) \varepsilon_1 \varepsilon_3\right] + k_\rho^2 k_x^2 \varepsilon_1 \\
B_{12} &= k_x k_y \left[\varepsilon_1 k_\rho^2 + \varepsilon_3 k_z^2\right] - k_0^2 \left[k_x k_y (\varepsilon_1^2 - \varepsilon_2^2) - i \varepsilon_2 \varepsilon_3 k_z^2\right] \\
B_{13} &= k_x k_z \left[\varepsilon_1 k_\rho^2 + \varepsilon_3 k_z^2\right] - k_0^2 \left[k_x k_z \varepsilon_1 \varepsilon_3 + i k_z k_y \varepsilon_2 \varepsilon_3\right] \\
B_{21} &= k_x k_y \left[\varepsilon_1 k_\rho^2 + \varepsilon_3 k_z^2\right] - k_0^2 \left[k_x k_y (\varepsilon_1^2 - \varepsilon_2^2) + i \varepsilon_2 \varepsilon_3 k_z^2\right] \\
B_{22} &= k_z^2 k_y^2 \varepsilon_3 + k_0^4 \varepsilon_3 (\varepsilon_1^2 - \varepsilon_2^2) - k_0^2 \left[k_y^2 (\varepsilon_1^2 - \varepsilon_2^2) + (k_\rho^2 + k_z^2) \varepsilon_1 \varepsilon_3\right] + k_\rho^2 k_y^2 \varepsilon_1 \\
B_{23} &= k_z k_y \left[\varepsilon_1 k_\rho^2 + \varepsilon_3 k_z^2\right] - k_0^2 \left[k_z k_y \varepsilon_1 \varepsilon_3 - i k_x k_z \varepsilon_2 \varepsilon_3\right] \\
B_{31} &= k_x k_z \left[\varepsilon_1 k_\rho^2 + \varepsilon_3 k_z^2\right] - k_0^2 \left[k_x k_z \varepsilon_1 \varepsilon_3 - i k_z k_y \varepsilon_2 \varepsilon_3\right] \\
B_{32} &= k_z k_y \left[\varepsilon_1 k_\rho^2 + \varepsilon_3 k_z^2\right] - k_0^2 \left[k_z k_y \varepsilon_1 \varepsilon_3 + i k_x k_z \varepsilon_2 \varepsilon_3\right] \\
B_{33} &= k_z^4 \varepsilon_3 + k_0^4 \varepsilon_3 (\varepsilon_1^2 - \varepsilon_2^2) - k_0^2 \left[k_\rho^2 (\varepsilon_1^2 - \varepsilon_2^2) + 2 k_z^2 \varepsilon_1 \varepsilon_3\right] + k_\rho^2 k_z^2 \varepsilon_1
\end{aligned} \tag{4.128}$$

We perform the integration over k_z after substituting (4.97), (4.127) into (4.122). The poles of the integrand occur at the zeros of $|\bar{\bar{W}}_E|$ denoted by $k_z = \pm k_{zI}$ and $k_z = \pm k_{zII}$ where k_{zI} and k_{zII} are defined by (3.62) and (3.63). Assuming the medium to be slightly lossy, i.e. $\mathrm{Im}k_z << \mathrm{Re}k_z, \mathrm{Im}k_z > 0$ and performing the contour integration over k_z, we obtain $\bar{\bar{G}}_{mm}(\bar{r},\bar{r}')$ in the following dyadic form for z > z':

$$\bar{\bar{G}}_{mm}(\bar{r},\bar{r}') = \frac{-\omega\varepsilon_0}{8\pi^2}\int_{-\infty}^{\infty}\int_{-\infty}^{\infty} dk_x dk_y \left\{ \frac{1}{k_0{}^2\varepsilon_3(k_{zI}{}^2 - k_{zII}{}^2)} \left[\frac{\beta_I}{k_{zI}}\hat{h}_{nI}(k_{zI})\hat{h}^*_{nI}(k_{zI})e^{i\bar{k}_I\cdot(\bar{r}-\bar{r}')} - \frac{\beta_{II}}{k_{zII}}\hat{h}_{nII}(k_{zII})\hat{h}^*_{nII}(k_{zII})e^{i\bar{k}_{II}\cdot(\bar{r}-\bar{r}')} \right] \right\}, \quad z>z' \tag{4.129a}$$

Similarly, when z < z', $\bar{\bar{G}}_{mm}(\bar{r},\bar{r}')$ can be obtained by assuming $\mathrm{Im}(-k_{zI}) < 0$ and $\mathrm{Im}(-k_{zII}) < 0$ and can be expressed as

$$\bar{\bar{G}}^e_{mm}(\bar{r},\bar{r}') = \frac{-\omega\varepsilon_0}{8\pi^2}\int_{-\infty}^{\infty}\int_{-\infty}^{\infty} dk_x dk_y \left\{ \frac{1}{k_0{}^2\varepsilon_3(k_{zI}{}^2 - k_{zII}{}^2)} \left[\frac{\beta_I}{k_{zI}}\hat{h}_{nI}(-k_{zI})\hat{h}^*_{nI}(-k_{zI})e^{i\bar{K}_I\cdot(\bar{r}-\bar{r}')} - \frac{\beta_{II}}{k_{zII}}\hat{h}_{nII}(-k_{zII})\hat{h}^*_{nII}(-k_{zII})e^{i\bar{K}_{II}\cdot(\bar{r}-\bar{r}')} \right] \right\}, \quad z<z' \tag{4.129b}$$

$\bar{k}_I$, $\bar{k}_{II}$, $\bar{K}_I$ and $\bar{K}_{II}$ are defined by (4.102). $\hat{h}_{nI}(\pm k_{zI})$ and $\hat{h}_{nII}(\pm k_{zI})$ are the orthonormal eigenvectors representing two characteristic *magnetic* fields for the type I and type II waves that exist in a gyroelectric medium and defined as

$$\hat{h}_{nI}(\pm k_{zI}) = \frac{\bar{h}_I(\pm k_{zI})}{norm(\bar{h}_I(\pm k_{zI}))} \tag{4.130a}$$

$$\hat{h}_{nII}(\pm k_{zII}) = \frac{\bar{h}_{II}(\pm k_{zII})}{norm(\bar{h}_{II}(\pm k_{zII}))} \tag{4.130b}$$

where

$$\bar{h}_I(\pm k_{zI}) = \begin{bmatrix} 1 \\ \dfrac{B_{13}B_{21} + B_{23}\beta_I - B_{23}B_{11}}{\beta_I B_{13} - B_{22}B_{13} + B_{23}B_{12}} \\ \dfrac{-B_{12}}{B_{13}}\left[\dfrac{B_{13}B_{21} + B_{23}\beta_I - B_{23}B_{11}}{\beta_I B_{13} - B_{22}B_{13} + B_{13}B_{12}}\right] + \dfrac{\beta_I - B_{11}}{B_{13}} \end{bmatrix} \tag{4.131a}$$

$$\bar{h}_{II}(\pm k_{zI}) = \begin{bmatrix} 1 \\ \dfrac{B_{13}B_{21} + B_{23}\beta_{II} - B_{23}B_{11}}{\beta_{II}B_{13} - B_{22}B_{13} + B_{23}B_{12}} \\ \dfrac{-B_{12}}{B_{13}}\left[\dfrac{B_{13}B_{21} + B_{23}\beta_{II} - B_{23}B_{11}}{\beta_{II}B_{13} - B_{22}B_{13} + B_{13}B_{12}}\right] + \dfrac{\beta_{II} - B_{11}}{B_{13}} \end{bmatrix} \tag{4.131b}$$

β_I and β_{II} are the corresponding eigenvalues for $\hat{h}_{nI}(\pm k_{zI})$ and $\hat{h}_{nII}(\pm k_{zI})$, respectively. They are defined as

$$\beta_I = k_I^4\left[\varepsilon_3\cos^2\theta + \varepsilon_1\sin^2\theta\right] - 2k_I^2{k_0}^2 \\ \times\left[\varepsilon_1\varepsilon_3\left(1+\cos^2\theta\right) + \left(\varepsilon_1^2 - \varepsilon_2^2\right)\sin^2\theta\right] + 3{k_0}^4\varepsilon_3\left({\varepsilon_1}^2 - {\varepsilon_2}^2\right) \tag{4.132a}$$

$$\beta_{II} = k_{II}^4\left[\varepsilon_3\cos^2\theta + \varepsilon_1\sin^2\theta\right] - 2k_{II}^2{k_0}^2 \\ \times\left[\varepsilon_1\varepsilon_3\left(1+\cos^2\theta\right) + \left(\varepsilon_1^2 - \varepsilon_2^2\right)\sin^2\theta\right] + 3{k_0}^4\varepsilon_3\left({\varepsilon_1}^2 - {\varepsilon_2}^2\right) \tag{4.132b}$$

Electric-magnetic type DGF, $\bar{\bar{G}}_{em}(\bar{r},\bar{r}')$, magnetic-electric type DGF, $\bar{\bar{G}}_{me}(\bar{r},\bar{r}')$ can now be easily obtained from the first order dyadic differential equations given by (4.27) and (4.25), respectively.

4.8 Application of Duality Principle

In this section we apply duality transformation that is outlined in Sect. 4.3 to develop the complete set of the DGFs for a magnetically gyrotropic or gyromagnetic medium. The duality relations between the DGFs of gyromagnetic and gyroelectric medium are obtained as

$$\bar{\bar{\varepsilon}} \rightarrow \bar{\bar{\mu}}, \quad \mu \rightarrow \varepsilon, \quad \mu_0 \rightarrow \varepsilon_0 \tag{4.133}$$

on the problem of electrically gyrotropic medium transforms dyadic Green's functions into their dual ones for a magnetically gyrotropic medium as follows.

$$\bar{\bar{G}}^e_{ee}(\bar{r},\bar{r}') \rightarrow \bar{\bar{G}}^m_{mm}(\bar{r},\bar{r}') \tag{4.134}$$

$$\bar{\bar{G}}^e_{me}(\bar{r},\bar{r}') \rightarrow -\bar{\bar{G}}^m_{em}(\bar{r},\bar{r}') \tag{4.135}$$

$$\bar{\bar{G}}^e_{mm}(\bar{r},\bar{r}') \rightarrow \bar{\bar{G}}^m_{ee}(\bar{r},\bar{r}') \tag{4.136}$$

$$\bar{\bar{G}}^{e}_{em}(\bar{r},\bar{r}') \rightarrow -\bar{\bar{G}}^{m}_{me}(\bar{r},\bar{r}') \tag{4.137}$$

At this point we introduce superscript e,m to distinguish DGFs for gyromagnetic and gyroelectric medium. The superscript 'e' stands for an electrically gyrotropic medium. The superscript 'm' stands for a magnetically gyrotropic medium.

4.8.1 Electric Type DGF $\bar{\bar{G}}_{ee}(\bar{r},\bar{r}')$ for a Gyromagnetic Medium

The electric type DGF $\bar{\bar{G}}^{m}_{ee}(\bar{r},\bar{r}')$ for a magnetically gyrotropic medium can be found by applying the duality transformation given by (4.133) and (4.136) as

$$\bar{\bar{\varepsilon}} \rightarrow \bar{\bar{\mu}}, \quad \mu \rightarrow \varepsilon, \quad \mu_0 \rightarrow \varepsilon_0$$

and

$$\bar{\bar{G}}^{e}_{mm}(\bar{r},\bar{r}') \rightarrow \bar{\bar{G}}^{m}_{ee}(\bar{r},\bar{r}')$$

Hence, we obtain the following result for $\bar{\bar{G}}^{m}_{ee}(\bar{r},\bar{r}')$ when z > z':

$$\bar{\bar{G}}^{m}_{ee}(\bar{r},\bar{r}') = \frac{-\omega\mu_0}{8\pi^2} \underset{z>z'}{\int_{-\infty}^{\infty}\int_{-\infty}^{\infty}} dk_x dk_y \left\{ \frac{1}{k_0{}^2\mu_3(k_{zI}'^2 - k_{zII}'^2)} \left[\frac{\alpha_I'}{k_{zI}'} \hat{e}_{nI}'(k_{zI}')\hat{e}_{nI}'^{*}(k_{zI}') e^{i\bar{k}_I'\cdot(\bar{r}-\bar{r}')} - \frac{\alpha_{II}'}{k_{zII}'} \hat{e}_{nII}'(k_{zII}')\hat{e}_{nII}'^{*}(k_{zII}') e^{i\bar{k}_{II}'\cdot(\bar{r}-\bar{r}')} \right] \right\}, \tag{4.138a}$$

Similarly, the DGF $\bar{\bar{G}}^{m}_{ee}(\bar{r},\bar{r}')$ when z < z' can be found as

$$\bar{\bar{G}}^{m}_{ee}(\bar{r},\bar{r}') = \frac{-\omega\mu_0}{8\pi^2} \underset{z<z'}{\int_{-\infty}^{\infty}\int_{-\infty}^{\infty}} dk_x dk_y \left\{ \frac{1}{k_0{}^2\mu_3(k_{zI}'^2 - k_{zII}'^2)} \left[\frac{\alpha_I'}{k_{zI}'} \hat{e}_{nI}'(-k_{zI}')\hat{e}_{nI}'^{*}(-k_{zI}') e^{i\bar{\kappa}_I'\cdot(\bar{r}-\bar{r}')} - \frac{\alpha_{II}'}{k_{zII}'} \hat{e}_{nII}'(-k_{zII}')\hat{e}_{nII}'^{*}(-k_{zII}') e^{i\bar{\kappa}_{II}'\cdot(\bar{r}-\bar{r}')} \right] \right\}, \tag{4.138b}$$

with the application of following transformations

$$\begin{bmatrix} k_{zi} \\ \bar{k}_i \\ \bar{\kappa}_i \\ \beta_i \\ \hat{h}_{ni} \end{bmatrix} \rightarrow \begin{bmatrix} k'_{zi} \\ \bar{k}'_i \\ \bar{\kappa}'_j \\ \alpha'_i \\ \hat{e}'_{ni} \end{bmatrix}_{i=I,II} \tag{4.139}$$

where

$$\begin{bmatrix} \varepsilon_1 & -i\varepsilon_2 & 0 \\ i\varepsilon_2 & \varepsilon_1 & 0 \\ 0 & 0 & \varepsilon_3 \end{bmatrix} \rightarrow \begin{bmatrix} \mu_1 & -i\mu_2 & 0 \\ i\mu_2 & \mu_1 & 0 \\ 0 & 0 & \mu_3 \end{bmatrix},$$

$$\mu \rightarrow \varepsilon, \quad \varepsilon_0 \rightarrow \mu_0$$

4.8.2 Magnetic Type DGF $\bar{\bar{G}}^m_{mm}(\bar{r},\bar{r}')$ for a Gyromagnetic Medium

We can find the magnetic type DGF $\bar{\bar{G}}^m_{mm}(\bar{r},\bar{r}')$ for a magnetically gyrotropic medium when we apply the following duality transformation given by (4.133) and (4.134) as

$$\bar{\bar{\varepsilon}} \rightarrow \bar{\bar{\mu}}, \quad \mu \rightarrow \varepsilon, \quad \mu_0 \rightarrow \varepsilon_0$$

$$\bar{\bar{G}}^e_{ee}(\bar{r},\bar{r}') \rightarrow \bar{\bar{G}}^m_{mm}(\bar{r},\bar{r}')$$

As a result, we obtain $\bar{\bar{G}}^m_{mm}(\bar{r},\bar{r}')$ for z > z′ as

$$\bar{\bar{G}}^m_{mm}(\bar{r},\bar{r}') = \frac{-\omega\varepsilon_0}{8\pi^2}\int\limits_{-\infty}^{\infty}\int\limits_{-\infty}^{\infty} dk_x dk_y \left\{ \frac{1}{k_0{}^2\mu_3(k'^2_{zI}-k'^2_{zII})}\left[\frac{\beta'_I}{k'_{zI}}\tilde{h}'_{nI}(k'_{zI})\tilde{h}'^*_{nI}(k'_{zI})e^{i\bar{k}'_I\cdot(\bar{r}-\bar{r}')} - \frac{\beta'_{II}}{k'_{zII}}\tilde{h}'_{nII}(k'_{zII})\tilde{h}'^*_{nII}(k'_{zII})e^{i\bar{k}'_{II}\cdot(\bar{r}-\bar{r}')}\right]\right\}, \quad z>z' \tag{4.140a}$$

Similarly, the DGF $\bar{\bar{G}}^m_{mm}(\bar{r},\bar{r}')$ when z < z′ is found as

$$\bar{\bar{G}}^m_{mm}(\bar{r},\bar{r}') = \frac{-\omega\varepsilon_0}{8\pi^2}\int_{-\infty}^{\infty}\int_{-\infty}^{\infty} dk_x dk_y \left\{ \begin{array}{l} \dfrac{1}{k_0{}^2\mu_3(k'^2_{zI}-k'^2_{zII})}\left[\dfrac{\beta'_I}{k'_{zI}}\hat{h}'_{nI}(-k'_{zI})\hat{h}'^*_{nI}(-k'_{zI})e^{i\bar{\kappa}'_I\cdot(\bar{r}-\bar{r}')} \right. \\ \left. \qquad -\dfrac{\beta'_{II}}{k'_{zII}}\hat{h}'_{nII}(-k'_{zII})\hat{h}'^*_{nII}(-k'_{zII})e^{i\bar{\kappa}'_{II}\cdot(\bar{r}-\bar{r}')}\right] \\ z<z' \end{array} \right\}, \tag{4.140b}$$

with the application of following transformations

$$\begin{bmatrix} k_{zi} \\ \bar{k}_i \\ \bar{\kappa}_i \\ \alpha_i \\ \hat{e}_{ni} \end{bmatrix} \rightarrow \begin{bmatrix} k'_{zi} \\ \bar{k}'_i \\ \bar{\kappa}'_i \\ \beta'_i \\ \hat{h}'_{ni} \end{bmatrix}_{i=I,II} \tag{4.141}$$

where

$$\begin{bmatrix} \varepsilon_1 & -i\varepsilon_2 & 0 \\ i\varepsilon_2 & \varepsilon_1 & 0 \\ 0 & 0 & \varepsilon_3 \end{bmatrix} \rightarrow \begin{bmatrix} \mu_1 & -i\mu_2 & 0 \\ i\mu_2 & \mu_1 & 0 \\ 0 & 0 & \mu_3 \end{bmatrix},$$

$$\mu \rightarrow \varepsilon, \quad \varepsilon_0 \rightarrow \mu_0$$

Similarly, electric-magnetic type DGF, $\bar{\bar{G}}^m_{em}(\bar{r},\bar{r}')$, magnetic-electric type DGF, $\bar{\bar{G}}^m_{me}(\bar{r},\bar{r}')$ for gyromagnetic medium can now be obtained from the first order dyadic differential equations.

References

1. Lee, J.K., and Kong, J.A., "Dyadic Green's functions for layered anisotropic medium," *Electromagnetics*, 3, 111–130, 1983.
2. R. Courant and D. Hilbert, *Methods of Mathematical Physics,* Vol. 1, Interscience Publishers, New York, Second Printing 1955.
3. A. Eroglu, *Electromagnetic Wave Propagation and Radiation in Gyrotropic Medium*, Ph.D. dissertation, Dept. of Electrical Eng. and Computer Science, Syracuse University, 2004.
4. Y.H. Lee, *Microwave Remote Sensing of Multilayered Anisotropic Random Media*, Dept. of Electrical Eng. and Computer Science, Syracuse University, 1993.

Chapter 5
Radiation in Anisotropic Medium

In this chapter, we will discuss the radiation characteristic of anisotropic medium by studying specifically dipole radiation from a uniaxially layered anisotropic media. We will calculate the far field radiation from an arbitrarily oriented Hertzian dipole when the dipole is placed over or embedded in a layered uniaxially anisotropic medium which is bounded above and below by isotropic media. The optic axis of the uniaxially anisotropic medium is arbitrarily oriented, i.e., not necessarily perpendicular to the plane of stratification. This leads to the cross-polarization effect and coupling between the ordinary and the extraordinary waves that exist in the anisotropic layer. The spectral domain approach is used to determine the far field behavior of the dipole. For this purpose, the dyadic Green's functions (DGFs) derived in Chap. 4 will be used. The far-field approximated Green's functions are evaluated using the method of stationary phase and the analytical results for the radiation fields are obtained for both horizontal ($\hat{x}-$ and $\hat{y}-$ oriented) and vertical ($\hat{z}-$ oriented) dipoles. The physical interpretation of the analytical results will also be discussed. In Sect. 5.5, numerical results are presented, including parameter studies on radiation patterns. In particular, the effects of anisotropy, layer thickness and dipole location on the radiation fields are discussed.

5.1 Formulation of the Problem

The anisotropic medium under consideration is *uniaxial* and characterized by a permitivity tensor $\bar{\bar{\varepsilon}}$ of the following form when its optic axis is along the z-axis

$$\bar{\bar{\varepsilon}}_1{}^{(o)} = \begin{bmatrix} \varepsilon_1 & 0 & 0 \\ 0 & \varepsilon_1 & 0 \\ 0 & 0 & \varepsilon_{1z} \end{bmatrix} \tag{5.1}$$

A. Eroglu, *Wave Propagation and Radiation in Gyrotropic and Anisotropic Media*,
DOI 10.1007/978-1-4419-6024-5_5, © Springer Science+Business Media, LLC 2010

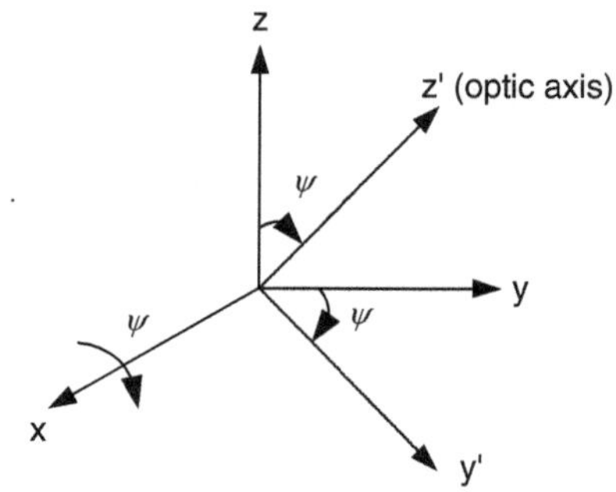

Fig. 5.1 Geometry of a uniaxially anisotropic medium

The medium is magnetically isotropic with permeability $\mu_0 = 4\pi \times 10^{-7}$ H/m. When the optic axis of the anisotropic medium is tilted off the z-axis by angle Ψ on the yz plane as shown in Fig. 5.1, its permitivity tensor is transformed accordingly as

$$\bar{\bar{\varepsilon}}_1 = \begin{bmatrix} \varepsilon_{11} & 0 & 0 \\ 0 & \varepsilon_{22} & \varepsilon_{23} \\ 0 & \varepsilon_{32} & \varepsilon_{33} \end{bmatrix} \tag{5.2}$$

where

$$\begin{aligned} \varepsilon_{11} &= \varepsilon_1 \\ \varepsilon_{22} &= \varepsilon_1 \cos^2\psi + \varepsilon_{1z} \sin^2\psi \\ \varepsilon_{23} &= \varepsilon_{32} = (\varepsilon_{1z} - \varepsilon_1) \cos\psi \sin\psi \\ \varepsilon_{33} &= \varepsilon_1 \sin^2\psi + \varepsilon_{1z} \cos^2\psi \end{aligned} \tag{5.3}$$

We first consider the problem of two-layered stratified medium as shown in Fig. 5.2. The upper and lower media are isotropic and characterized by permitivities ε_0 and ε_2, respectively. The medium in the middle is uniaxially anisotropic as described above. The dyadic Green's functions (DGF) of this structure when the unit impulse source is located at $z = z'$ in region 0, i.e., $z' > 0$ satisfy

$$\nabla \times \nabla \times \bar{\bar{G}}_{00}(\bar{r}, \bar{r}') - \omega^2 \mu_0 \varepsilon_0 \bar{\bar{G}}_{00}(\bar{r}, \bar{r}') = \bar{\bar{I}} \delta(\bar{r} - \bar{r}'), \ z \geq 0 \tag{5.4a}$$

$$\nabla \times \nabla \times \bar{\bar{G}}_{10}(\bar{r}, \bar{r}') - \omega^2 \mu_0 \bar{\bar{\varepsilon}}_1 \cdot \bar{\bar{G}}_{10}(\bar{r}, \bar{r}') = 0, \ -d \leq z \leq 0 \tag{5.4b}$$

$$\nabla \times \nabla \times \bar{\bar{G}}_{20}(\bar{r}, \bar{r}') - \omega^2 \mu_0 \varepsilon_2 \bar{\bar{G}}_{20}(\bar{r}, \bar{r}') = 0, \ z \leq -d \tag{5.4c}$$

with the boundary conditions at $z = 0$ and $z = -d$ and radiation condition at $z = \pm\infty$. The first subscript of the DGFs refers to the region of the field point

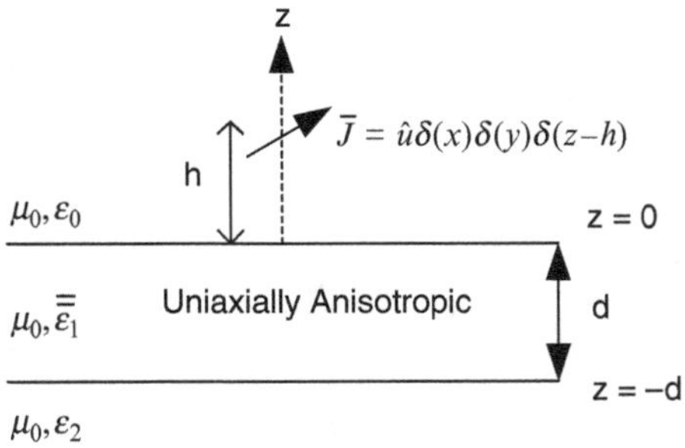

Fig. 5.2 Geometry of the problem when the dipole is over two-layered anisotropic medium

and the second subscript to the region containing the source. Explicit solutions of DGFs $\bar{\bar{G}}_{00}(\bar{r}, \bar{r}')$ for z < z', $\bar{\bar{G}}_{10}(\bar{r}, \bar{r}')$ and $\bar{\bar{G}}_{20}(\bar{r}, \bar{r}')$ are obtained by Lee and Kong [1] when the source is in region 0.

5.2 Far Field Radiation: Dipole Is Over Layered Uniaxially Anisotropic Media

In order to determine the far field radiation from a dipole over a two-layered anisotropic medium, the far field approximated DGF of the problem in Fig. 5.2 has to be found. In particular, we need to find the DGF $\bar{\bar{G}}_{00}(\bar{r}, \bar{r}')$ of the problem when $z > z'$. Noting that the electric field $\bar{E}_0$ in region 0 satisfies

$$\nabla \times \nabla \times \bar{E}_0(\bar{r}) - \omega^2 \mu_0 \varepsilon_0 \bar{E}_0(\bar{r}) = i\omega\mu_0 \bar{J}(\bar{r}) \tag{5.5}$$

the radiation field can be calculated using the DGF as follows.

$$\bar{E}_0(r) = i\omega\mu_0 \iiint \bar{\bar{G}}_{00}(\bar{r}, \bar{r}') \cdot \bar{J}(\bar{r}') dV' \tag{5.6}$$

where

$$\bar{J}(\bar{r}') = \hat{u} c \delta(x')\delta(y')\delta(z' - h) \tag{5.7}$$

when the source is the Hertzian dipole. In (5.7), c is the current moment of the electric dipole and $\hat{u}$ is the orientation of the dipole. After evaluating the integral in (5.6), it will be possible to express $\bar{E}_0(\bar{r})$ in terms of horizontally polarized and vertically polarized components as

$$\bar{E}_0(r) = \hat{h}(k_{oz}) E_h(r) + \hat{v}(k_{oz}) E_v(r) \tag{5.8}$$

The DGF $\bar{\bar{G}}_{00}(\bar{r},\bar{r}')$ when z < z′ is obtained in [1] and given by

$$\bar{\bar{G}}_{00}(\bar{r},\bar{r}') = \frac{i}{8\pi^2}\int\int_{-\infty}^{\infty} dk_x dk_y \frac{1}{k_{0z}}\left\{\hat{h}(-k_{0z})\left[\hat{h}(-k_{0z})e^{i(\bar{\kappa}_0\cdot\bar{r})} + R_{HH}\hat{h}(k_{0z})e^{i(\bar{k}_0\cdot\bar{r})}\right.\right.$$
$$\left.+R_{HV}\hat{v}(k_{0z})e^{i(\bar{k}_0\cdot\bar{r})}\right]. + \hat{v}(-k_{0z})\left[\hat{v}(-k_{0z})e^{i(\bar{\kappa}_0\cdot\bar{r})} + R_{VV}\hat{v}(k_{0z})e^{i(\bar{k}_0\cdot\bar{r})}\right.$$
$$\left.\left.+R_{VH}\hat{h}(k_{0z})e^{i(\bar{k}_0\cdot\bar{r})}\right]\right\}e^{-i(\bar{\kappa}_0\cdot\bar{r}')}, \quad z<z' \tag{5.9a}$$

To derive the DGF $\bar{\bar{G}}_{00}(\bar{r},\bar{r}')$ when z > z′, we employ the symmetric property [2] of the DGF $\bar{\bar{G}}_{00}(\bar{r},\bar{r}')$, and make use of the DGF $\bar{\bar{G}}_{00}(\bar{r},\bar{r}')$ for z < z′, as follows.

$$\bar{\bar{G}}_{00}(\bar{r},\bar{r}')\Big|_{z>z'} = \left[\bar{\bar{G}}_{00}(\bar{r}',\bar{r})\right]^T\Big|_{z<z'} \tag{5.9b}$$

where the superscript T stands for the transpose. The new DGF for z > z′ is given by

$$\bar{\bar{G}}_{00}(\bar{r},\bar{r}')=\frac{i}{8\pi^2}\int\int_{-\infty}^{\infty} dk_x dk_y\frac{1}{k_{0z}}\left\{\hat{h}(k_{0z})\left[\hat{h}(k_{0z})e^{-i(\bar{k}_0\cdot\bar{r}')}+R'_{HH}\hat{h}(-k_{0z})e^{-i(\bar{\kappa}_0\cdot\bar{r}')}\right.\right.$$
$$\left.+R'_{HV}\hat{v}(-k_{0z})e^{-i(\bar{\kappa}_0\cdot\bar{r}')}\right]+\hat{v}(k_{0z})\left[\left[\hat{v}.(k_{0z})e^{-i(\bar{k}_0\cdot\bar{r}')}+R'_{VV}\hat{v}(-k_{0z})e^{-i(\bar{\kappa}_0\cdot\bar{r}')}\right.\right.$$
$$\left.\left.+R'_{VH}\hat{h}(-k_{0z})e^{-i(\bar{\kappa}_0\cdot\bar{r}')}\right]\right\}e^{i(\bar{k}_0\cdot\bar{r})}, \quad z>z' \tag{5.10}$$

where

$$\begin{aligned} R'_{HH} &= R_{HH}(k_x \to -k_x, k_y \to -k_y) \\ R'_{HV} &= -R_{HV}(k_x \to -k_x, k_y \to -k_y) \\ R'_{VV} &= R_{VV}(k_x \to -k_x, k_y \to -k_y) \\ R'_{VH} &= -R_{VH}(k_x \to -k_x, k_y \to -k_y) \end{aligned} \tag{5.11}$$

and all the vectors and coefficients including R_{HH}, R_{HV}, R_{VV} and R_{VH} are defined in Appendix. We note that $\hat{h}(\pm k_{oz})$ is a unit vector in the direction of an electric field for horizontally polarized wave and $\hat{v}(\pm k_{oz})$ is a unit vector for vertically polarized wave where (+) refers to upward propagating wave and (−) to downward propagating wave. Now under the far field approximation, the integral for $\bar{\bar{G}}_{00}(\bar{r},\bar{r}')$ in (5.10) is evaluated with the method of stationary phase [3, 4] as r → ∞. The result is

$$\bar{\bar{G}}_{00}(\bar{r},\bar{r}') = \frac{e^{ik_0 r}}{4\pi r}\bar{\bar{g}}_0(\bar{k}_\rho,\bar{r}') \tag{5.12}$$

where

$$\begin{aligned}\bar{\bar{g}}_0(\bar{k}_\rho,\bar{r}') = {} & \hat{h}(k_{0z})\left[\hat{h}(k_{0z})e^{-i(\bar{k}_0\cdot\bar{r}')} + R'_{HH}\hat{h}(-k_{0z})e^{-i(\bar{\kappa}_0\cdot\bar{r}')} + R'_{HV}\hat{v}(-k_{0z})e^{-i(\bar{\kappa}_0\cdot\bar{r}')}\right] \\ & + \hat{v}(k_{0z})\left[\hat{v}(k_{0z})e^{-i(\bar{k}_0\cdot\bar{r}')} + R'_{VV}\hat{v}(-k_{0z})e^{-i(\bar{\kappa}_0\cdot\bar{r}')} + R'_{VH}\hat{h}(-k_{0z})e^{-i(\bar{\kappa}_0\cdot\bar{r}')}\right]\end{aligned} \tag{5.13}$$

$$\begin{aligned}\bar{k}_0 &= \bar{k}_\rho + \hat{z}k_{0z} \\ \bar{\kappa}_0 &= \bar{k}_\rho - \hat{z}k_{0z} \\ \bar{k}_\rho &= \hat{x}k_x + \hat{y}k_y = \hat{x}k_0\sin\theta\cos\varphi + \hat{y}k_0\sin\theta\sin\varphi\end{aligned} \tag{5.14}$$

and $(\theta,\ \phi)$ are observation angles at the field point $\bar{r}$. Radiation fields can now be found by substituting (5.12) into (5.6). The result is

$$\begin{aligned}\bar{E}_0(\bar{r}) &= i\omega\mu_o\bar{\bar{G}}_{00}(\bar{r},\bar{r}')\Big|_{x'=0,\ y'=0,\ z'=h} \bullet\hat{u}c \\ &= i\omega\mu_0 c\frac{e^{ik_o r}}{4\pi r}\bar{\bar{g}}_0(\bar{k}_\rho,\bar{r}')\Big|_{x'=0,\ y'=0,\ z'=h} \bullet\hat{u}\end{aligned} \tag{5.15}$$

When the dipole is oriented in $\hat{z}$ direction, $\hat{u}=\hat{z}$, (5.15) can be written in the form of (5.8) where

$$E_h(r) = \frac{\omega\mu_0 c}{i}\frac{e^{i\bar{k}_0\cdot\bar{r}}}{4\pi r}\left\{\frac{k_\rho}{k_0}R'_{HV}e^{ik_{0z}d}\right\} \tag{5.16a}$$

$$E_v(r) = \frac{\omega\mu_0 c}{i}\frac{e^{i\bar{k}_o\cdot\bar{r}}}{4\pi r}\left\{\frac{k_\rho}{k_o}(e^{-ik_{oz}d} + R'_{VV}e^{ik_{oz}d})\right\} \tag{5.16b}$$

When the dipole is oriented in $\hat{x}$ direction, $\hat{u}=\hat{x}$, we obtain

$$E_h(r) = \frac{\omega\mu_0 c}{i}\frac{e^{i\bar{k}_o\cdot\bar{r}}}{4\pi r}\left\{\frac{k_y}{k_\rho}(e^{-ik_{oz}d} + R'_{HH}e^{ik_{oz}d}) + \frac{k_x k_{oz}}{k_\rho k_o}R'_{HV}e^{ik_{oz}d}\right\} \tag{5.17a}$$

$$E_v(r) = \frac{\omega\mu_0 c}{i}\frac{e^{i\bar{k}_o.\bar{r}}}{4\pi r}\left\{-\frac{k_x k_{oz}}{k_\rho k_o}(e^{-ik_{oz}d} - R'_{VV}e^{ik_{oz}d}) + \frac{k_y}{k_\rho}R'_{VH}e^{ik_{oz}d}\right\} \tag{5.17b}$$

When the dipole is oriented in $\hat{y}$ direction, $\hat{u} = \hat{y}$, we obtain

$$E_h(r) = \frac{\omega\mu_0 c}{i}\frac{e^{i\bar{k}_o.\bar{r}}}{4\pi r}\left\{-\frac{k_x}{k_\rho}(e^{-ik_{oz}d} + R'_{HH}e^{ik_{oz}d}) + \frac{k_y k_{oz}}{k_\rho k_o}R'_{HV}e^{ik_{oz}d}\right\} \tag{5.18a}$$

$$E_v(r) = \frac{\omega\mu_0 c}{i}\frac{e^{i\bar{k}_o.\bar{r}}}{4\pi r}\left\{-\frac{k_y k_{oz}}{k_\rho k_o}(e^{-ik_{oz}d} - R'_{VV}e^{ik_{oz}d}) - \frac{k_x}{k_\rho}R'_{VH}e^{ik_{oz}d}\right\} \tag{5.18b}$$

The formulas (5.16–5.18) give a complete set of radiation fields for arbitrarily oriented dipoles when the dipole is at a height h above the anisotropic layer. We also notice that in spherical coordinate system, $\hat{h}$ component of the far field corresponds to $\hat{\varphi}$ component, $\hat{v}$ component of the far field corresponds to $\hat{\theta}$ component.

5.3 Far Field Radiation: Dipole Is Embedded Inside Two-Layered Anisotropic Medium

In this section, we will find the radiation from an arbitrarily oriented Hertzian dipole when the dipole is embedded inside a uniaxially anisotropic medium as shown in Fig. 5.3. All other parameters considered in Sect. 5.2 remain unchanged. For this problem, we need to evaluate the far-field approximated dyadic Green's function $\bar{\bar{G}}_{01}(\bar{r},\bar{r}')$. Namely, the source point is in region 1 and the observation point is in region 0. When the impulse source is located at z = z′ in region 1 ($-d < z' < 0$), the DGFs $\bar{\bar{G}}_{01}, \bar{\bar{G}}_{11}$, G_{21} satisfy the following vector wave equations:

$$\nabla\times\nabla\times\bar{\bar{G}}_{01}(\bar{r},\bar{r}') - \omega^2\mu_0\varepsilon_0\bar{\bar{G}}_{01}(\bar{r},\bar{r}') = 0,\ z \geq 0 \tag{5.19a}$$

$$\nabla\times\nabla\times\bar{\bar{G}}_{11}(\bar{r},\bar{r}') - \omega^2\mu_0\bar{\bar{\varepsilon}}_1\cdot\bar{\bar{G}}_{11}(\bar{r},\bar{r}') = \bar{\bar{I}}\delta(\bar{r}-\bar{r}'),\ -d \leq z \leq 0 \tag{5.19b}$$

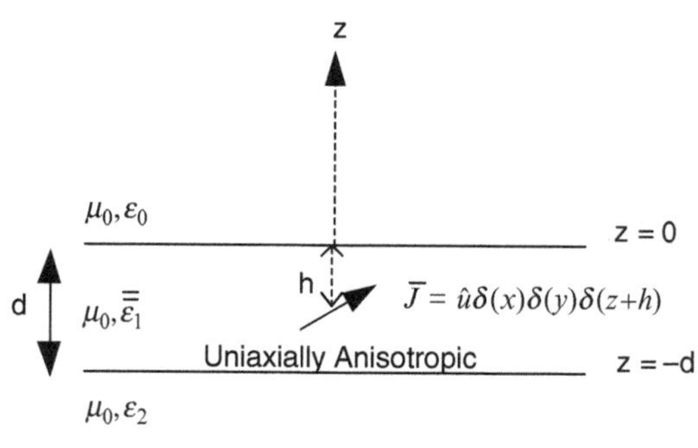

Fig. 5.3 Geometry of the problem when the dipole is embedded inside two-layered anisotropic medium

$$\nabla \times \nabla \times \bar{\bar{G}}_{21}(\bar{r},\bar{r}') - \omega^2\mu_0\varepsilon_2\bar{\bar{G}}_{21}(\bar{r},\bar{r}') = 0,\ z \le -d \tag{5.19c}$$

Instead of solving this new problem, we can again make the use of the results for the DGF $\bar{\bar{G}}_{10}(\bar{r},\bar{r}')$ derived in [1] and given by

$$\begin{aligned}\bar{\bar{G}}_{10}(\bar{r},\bar{r}') = \frac{i}{8\pi^2}\iint dk_x dk_y \frac{1}{k_{0z}} e^{-i\bar{\kappa}_o\cdot\bar{r}'}\Big\{\hat{h}(-k_{oz})\Big[A_{Ho}\hat{o}(-k_{1z}^o)e^{i(\bar{\kappa}_1^o\cdot\bar{r})}\\ + B_{Ho}\hat{o}(k_{1z}^o)e^{i(\bar{\kappa}_1^o\cdot\bar{r})} + A_{He}\hat{e}(k_{1z}^{ed})e^{i(\bar{\kappa}_1^e\cdot\bar{r})} + B_{He}\hat{e}(k_{1z}^{eu})e^{i(\bar{\kappa}_1^e\cdot\bar{r})}\Big]\\ + \hat{v}(-k_{oz})\Big[A_{Vo}\hat{o}(-k_{1z}^o)e^{i(\bar{\kappa}_1^o\cdot\bar{r})} + B_{Vo}\hat{o}(k_{1z}^o)e^{i(\bar{\kappa}_1^o\cdot\bar{r})} + A_{Ve}\hat{e}(k_{1z}^{ed})e^{i(\bar{\kappa}_1^e\cdot\bar{r})}\\ + B_{Ve}\hat{e}(k_{1z}^{eu})e^{i(\bar{\kappa}_1^e\cdot\bar{r})}\Big]\Big\},\ -d \le z \le 0\end{aligned} \tag{5.20a}$$

which is the DGF when the source point is in region 0 and the observation point in region 1. $\bar{\bar{G}}_{01}(\bar{r},\bar{r}')$ and $\bar{\bar{G}}_{10}(\bar{r},\bar{r}')$ satisfy the following symmetric property:

$$\bar{\bar{G}}_{01}(\bar{r},\bar{r}')\Big|_{z>0} = \Big[\bar{\bar{G}}_{10}(\bar{r}',\bar{r})\Big]^T\Big|_{z<0} \tag{5.20b}$$

The new DGF is then given by

$$\begin{aligned}\bar{\bar{G}}_{01}(\bar{r},\bar{r}') = \frac{i}{8\pi^2}\iint dk_x dk_y \frac{1}{k_{0z}} e^{i\bar{k}_o\cdot\bar{r}}\Big\{\hat{h}(k_{oz})\Big[A_{oH}\hat{o}(k_{1z}^o).e^{-i(\bar{k}_1^o\cdot\bar{r}')}\\ +B_{oH}\hat{o}(-k_{1z}^o)e^{-i(\bar{\kappa}_1^o\cdot\bar{r}')} + A_{eH}\hat{e}(k_{1z}^{eu})e^{-i(\bar{k}_1^e\cdot\bar{r}')} + B_{eH}\hat{e}(k_{1z}^{ed})e^{-i(\bar{\kappa}_1^e\cdot\bar{r}')}\Big]\\ + \hat{v}(k_{oz})\Big[A_{oV}\hat{o}(k_{1z}^o)e^{-i(k_1^o\cdot\bar{r}')} + B_{oV}\hat{o}(-k_{1z}^o)e^{-i(\bar{\kappa}_1^o\cdot\bar{r}')}\\ +A_{eV}\hat{e}(k_{1z}^{eu})e^{-i(\bar{k}_1^e\cdot\bar{r}')} + B_{eV}\hat{e}(k_{1z}^{ed})e^{-i(\bar{\kappa}_1^e\cdot\bar{r}')}\Big]\Big\},z > 0\end{aligned} \tag{5.21}$$

where

$$\begin{aligned}A_{oH} &= A_{Ho}(k_x \to -k_x, k_y \to -k_y)\\ B_{oH} &= B_{Ho}(k_x \to -k_x, k_y \to -k_y)\\ A_{eH} &= -A_{He}(k_x \to -k_x, k_y \to -k_y)\\ B_{eH} &= -B_{He}(k_x \to -k_x, k_y \to -k_y)\\ A_{oV} &= -A_{Vo}(k_x \to -k_x, k_y \to -k_y)\\ B_{oV} &= -B_{Vo}(k_x \to -k_x, k_y \to -k_y)\\ A_{eV} &= A_{Ve}(k_x \to -k_x, k_y \to -k_y)\\ B_{eV} &= B_{Ve}(k_x \to -k_x, k_y \to -k_y)\end{aligned} \tag{5.22}$$

and all the vectors, variables and coefficients including A_{Ho}, B_{Ho}, etc. are defined in Appendix. We note that $\hat{o}(\pm k_{1z}^{o})$ is a unit vector in the direction of an electric field for an ordinary wave and $\hat{e}(k_{1z}^{eu})$ or $\hat{e}(k_{1z}^{ed})$ is a unit vector for an extraordinary wave for an upward or downward propagating wave. Under the far field approximation, the integral for $\bar{\bar{G}}_{01}(\bar{r},\bar{r}')$ in (5.21) is evaluated again with the method of stationary phase. The result is

$$\bar{\bar{G}}_{01}(\bar{r},\bar{r}') = \frac{e^{ik_o r}}{4\pi r}\,\bar{\bar{g}}_1(\bar{k}_\rho,\bar{r}') \tag{5.23}$$

where

$$\begin{aligned}\bar{\bar{g}}_1(\bar{k}_\rho,\bar{r}') =& \hat{h}(k_{0z})\Big\{A_{oH}\hat{o}(k_{1z}^{o})e^{-i(\bar{k}_1^{o}\cdot\bar{r}')} + B_{oH}\hat{o}(-k_{1z}^{o})e^{-i(\bar{\kappa}_1^{o}\cdot\bar{r}')}\\ &+A_{eH}\hat{e}(k_{1z}^{eu})e^{-i(\bar{k}_1^{e}\cdot\bar{r}')} + B_{eH}\hat{e}(k_{1z}^{ed})e^{-i(\bar{\kappa}_1^{e}\cdot\bar{r}')}\Big\}\\ &+\hat{v}(k_{0z})\Big\{A_{oV}\hat{o}(k_{1z}^{o})e^{-i(k_1^{o}\cdot\bar{r}')} + B_{oV}\hat{o}(-k_{1z}^{o})e^{-i(\bar{\kappa}_1^{o}\cdot\bar{r}')}\\ &+A_{oV}\hat{o}(k_{1z}^{o})e^{-i(k_1^{o}\cdot\bar{r}')} + B_{oV}\hat{o}(-k_{1z}^{o})e^{-i(\bar{\kappa}_1^{o}\cdot\bar{r}')}\Big\}\end{aligned} \tag{5.24}$$

Radiation fields can be found by modifying (5.6) as follows:

$$\bar{E}_0(\bar{r}) = i\omega\mu_0\iiint\bar{\bar{G}}_{01}(\bar{r},\bar{r}')\bullet\bar{J}(\bar{r}')dV' = i\omega\mu_o\bar{\bar{G}}_{01}(\bar{r},\bar{r}')\Big|_{x'=0,\ y'=0,\ z'=-h}\bullet\hat{u}c \tag{5.25a}$$

Substituting (5.23) into (5.25a) and evaluating the integral, we obtain the radiation field:

$$\bar{E}_0(\bar{r}) = i\omega\mu_0 c\frac{e^{ik_o r}}{4\pi r}\bar{\bar{g}}_1(\bar{k}_\rho,\bar{r}')\Big|_{x'=0,\ y'=0,\ z'=-h}\bullet\hat{u} \tag{5.25b}$$

When the dipole is oriented in $\hat{z}$ direction, $\hat{u}=\hat{z}$, the components of the radiation fields are

$$\begin{aligned}E_h(r) =& \frac{\omega\mu_0 c}{i}\frac{e^{i\bar{k}_0\cdot\bar{r}}}{4\pi r}\Bigg\{\frac{A_{oH}(k_x\sin\psi)e^{ik_{1z}^{o}h}}{[(k_{1z}^{o}\sin\psi-k_y\cos\psi)^2+k_x^{\,2}]^{1/2}} + \frac{B_{oH}k_x\sin\psi e^{-ik_{1z}^{o}h}}{[(k_{1z}^{o}\sin\psi+k_y\cos\psi)^2+k_x^{\,2}]^{1/2}}\\ &-\frac{A_{eH}[(\varepsilon_{22}k_y+\varepsilon_{23}k_{1z}^{eu})(k_{1z}^{eu}\sin\psi-k_y\cos\psi)-k_x^{\,2}\cos\psi\varepsilon_{11}]e^{ik_{1z}^{eu}h}}{[(k_{1z}^{eu}\sin\psi-k_y\cos\psi)^2+k_x^{\,2}]^{1/2}[(k_x\varepsilon_{11})^2+(\varepsilon_{22}k_y+\varepsilon_{23}k_{1z}^{eu})^2+(\varepsilon_{32}k_y+\varepsilon_{33}k_{1z}^{eu})^2]^{1/2}}\\ &-\frac{B_eH[(\varepsilon_{22}k_y+\varepsilon_{23}k_{1z}^{ed})(k_{1z}^{ed}\sin\psi-k_y\cos\psi)-k_x^{\,2}\cos\psi\varepsilon_{11}]e^{ik_{1z}^{ed}h}}{[(k_{1z}^{ed}\sin\psi-k_y\cos\psi)^2+k_x^{\,2}]^{1/2}[(k_x\varepsilon_{11})^2+(\varepsilon_{22}k_y+\varepsilon_{23}k_{1z}^{ed})^2+(\varepsilon_{32}k_y+\varepsilon_{33}k_{1z}^{ed})^2]^{1/2}}\Bigg\}\end{aligned} \tag{5.26a}$$

and

$$E_v(r)=\frac{\omega\mu_0 c}{i}\frac{e^{i\bar{k}_0.\bar{r}}}{4\pi r}\left\{\frac{A_{oV}(k_x\sin\psi)e^{ik_{1z}^o h}}{[(k_{1z}^o\sin\psi-k_y\cos\psi)^2+k_x{}^2]^{1/2}}+\frac{B_{oV}k_x\sin\psi e^{-ik_{1z}^o h}}{[(k_{1z}^o\sin\psi+k_y\cos\psi)^2+k_x{}^2]^{1/2}}\right.$$
$$-\frac{A_{eV}[(\varepsilon_{22}k_y+\varepsilon_{23}k_{1z}^{eu})(k_{1z}^{eu}\sin\psi-k_y\cos\psi)-k_x{}^2\cos\psi\varepsilon_{11}]e^{ik_{1z}^{eu}h}}{[(k_{1z}^{eu}\sin\psi-k_y\cos\psi)^2+k_x{}^2]^{1/2}[(k_x\varepsilon_{11})^2+(\varepsilon_{22}k_y+\varepsilon_{23}k_{1z}^{eu})^2+(\varepsilon_{32}k_y+\varepsilon_{33}k_{1z}^{eu})^2]^{1/2}}$$
$$\left.-\frac{B_{eV}[(\varepsilon_{22}k_y+\varepsilon_{23}k_{1z}^{ed})(k_{1z}^{ed}\sin\psi-k_y\cos\psi)-k_x{}^2\cos\psi\varepsilon_{11}]e^{ik_{1z}^{ed}h}}{[(k_{1z}^{ed}\sin\psi-k_y\cos\psi)^2+k_x{}^2]^{1/2}[(k_x\varepsilon_{11})^2+(\varepsilon_{22}k_y+\varepsilon_{23}k_{1z}^{ed})^2+(\varepsilon_{32}k_y+\varepsilon_{33}k_{1z}^{ed})^2]^{1/2}}\right\}$$
$$(5.26b)$$

When the dipole is oriented in $\hat{x}$ direction, $\hat{u}=\hat{x}$,

$$E_h(r)=\frac{\omega\mu_0 c}{i}\frac{e^{i\bar{k}_0.\bar{r}}}{4\pi r}\left\{\frac{A_{oH}(k_y\cos\psi-k_{1z}^o\sin\psi)e^{ik_{1z}^o h}}{[(k_{1z}^o\sin\psi-k_y\cos\psi)^2+k_x{}^2]^{1/2}}+\frac{B_{oH}(k_{1z}^o\sin\psi+k_y\cos\psi)e^{-ik_{1z}^o h}}{[(k_{1z}^o\sin\psi+k_y\cos\psi)^2+k_x{}^2]^{1/2}}\right.$$
$$-\frac{A_{eH}[k_x\sin\psi(\varepsilon_{22}k_y+\varepsilon_{23}k_{1z}^{eu})+k_x\cos\psi(\varepsilon_{32}k_y+\varepsilon_{33}k_{1z}^{eu})]e^{ik_{1z}^{eu}h}}{[(k_{1z}^{eu}\sin\psi-k_y\cos\psi)^2+k_x{}^2]^{1/2}[(k_x\varepsilon_{11})^2+(\varepsilon_{22}k_y+\varepsilon_{23}k_{1z}^{eu})^2+(\varepsilon_{32}k_y+\varepsilon_{33}k_{1z}^{eu})^2]^{1/2}}$$
$$\left.-\frac{B_{eH}[k_x\sin\psi(\varepsilon_{22}k_y+\varepsilon_{23}k_{1z}^{ed})+k_x\cos\psi(\varepsilon_{32}k_y+\varepsilon_{33}k_{1z}^{ed})]e^{ik_{1z}^{ed}h}}{[(k_{1z}^{ed}\sin\psi-k_y\cos\psi)^2+k_x{}^2]^{1/2}[(k_x\varepsilon_{11})^2+(\varepsilon_{22}k_y+\varepsilon_{23}k_{1z}^{ed})^2+(\varepsilon_{32}k_y+\varepsilon_{33}k_{1z}^{ed})^2]^{1/2}}\right\}$$
$$(5.27a)$$

and

$$E_v(r)=\frac{\omega\mu_0 c}{i}\frac{e^{i\bar{k}_0.\bar{r}}}{4\pi r}\left\{\frac{A_{oV}(k_y\cos\psi-k_{1z}^o\sin\psi)e^{ik_{1z}^o h}}{[(k_{1z}^o\sin\psi-k_y\cos\psi)^2+k_x{}^2]^{1/2}}+\frac{B_{oV}(k_{1z}^o\sin\psi+k_y\cos\psi)e^{-ik_{1z}^o h}}{[(k_{1z}^o\sin\psi+k_y\cos\psi)^2+k_x{}^2]^{1/2}}\right.$$
$$-\frac{A_{eV}[k_x\sin\psi(\varepsilon_{22}k_y+\varepsilon_{23}k_{1z}^{eu})+k_x\cos\psi(\varepsilon_{32}k_y+\varepsilon_{33}k_{1z}^{eu})]e^{ik_{1z}^{eu}h}}{[(k_{1z}^{eu}\sin\psi-k_y\cos\psi)^2+k_x{}^2]^{1/2}[(k_x\varepsilon_{11})^2+(\varepsilon_{22}k_y+\varepsilon_{23}k_{1z}^{eu})^2+(\varepsilon_{32}k_y+\varepsilon_{33}k_{1z}^{eu})^2]^{1/2}}$$
$$\left.-\frac{B_{eV}[k_x\sin\psi(\varepsilon_{22}k_y+\varepsilon_{23}k_{1z}^{ed})+k_x\cos\psi(\varepsilon_{32}k_y+\varepsilon_{33}k_{1z}^{ed})]e^{ik_{1z}^{ed}h}}{[(k_{1z}^{ed}\sin\psi-k_y\cos\psi)^2+k_x{}^2]^{1/2}[(k_x\varepsilon_{11})^2+(\varepsilon_{22}k_y+\varepsilon_{23}k_{1z}^{ed})^2+(\varepsilon_{32}k_y+\varepsilon_{33}k_{1z}^{ed})^2]^{1/2}}\right\}$$
$$(5.27b)$$

When the dipole is oriented in $\hat{y}$ direction, $\hat{u} = \hat{y}$,

$$E_h(r) = \frac{\omega\mu_0 c}{i}\frac{e^{i\bar{k}_0.\bar{r}}}{4\pi r}\left\{\frac{-A_{oH}(k_x\cos\psi)e^{ik^o_{1z}h}}{[(k^o_{1z}\sin\psi - k_y\cos\psi)^2 + k_x^2]^{1/2}} + \frac{-B_{oH}(k_x\cos\psi)e^{-ik^o_{1z}h}}{[(k^o_{1z}\sin\psi + k_y\cos\psi)^2 + k_x^2]^{1/2}}\right.$$
$$+\frac{A_{eH}[(k^{eu}_{1z}\sin\psi - k_y\cos\psi)(\varepsilon_{32}k_y + \varepsilon_{23}k^{eu}_{1z}) + k_x^2\sin\psi\varepsilon_{11}]e^{ik^{eu}_{1z}h}}{[(k^{eu}_{1z}\sin\psi - k_y\cos\psi)^2 + k_x^2]^{1/2}[(k_x\varepsilon_{11})^2 + (\varepsilon_{22}k_y + \varepsilon_{23}k^{eu}_{1z})^2 + (\varepsilon_{32}k_y + \varepsilon_{33}k^{eu}_{1z})^2]^{1/2}}$$
$$\left.+\frac{B_{eH}[(k^{ed}_{1z}\sin\psi - k_y\cos\psi)(\varepsilon_{22}k_y + \varepsilon_{23}k^{ed}_{1z}) + k_x^2\sin\psi\varepsilon_{11}]e^{ik^{ed}_{1z}h}}{[(k^{ed}_{1z}\sin\psi - k_y\cos\psi)^2 + k_x^2]^{1/2}[(k_x\varepsilon_{11})^2 + (\varepsilon_{22}k_y + \varepsilon_{23}k^{ed}_{1z})^2 + (\varepsilon_{32}k_y + \varepsilon_{33}k^{ed}_{1z})^2]^{1/2}}\right\} \tag{5.28a}$$

and

$$E_v(r) = \frac{\omega\mu_0 c}{i}\frac{e^{i\bar{k}_0.\bar{r}}}{4\pi r}\left\{\frac{-A_{oV}(k_x\cos\psi)e^{ik^o_{1z}h}}{[(k^o_{1z}\sin\psi - k_y\cos\psi)^2 + k_x^2]^{1/2}} + \frac{-B_{oV}(k_x\cos\psi)e^{-ik^o_{1z}h}}{[(k^o_{1z}\sin\psi + k_y\cos\psi)^2 + k_x^2]^{1/2}}\right.$$
$$+\frac{A_{eV}[(k^{eu}_{1z}\sin\psi - k_y\cos\psi)(\varepsilon_{32}k_y + \varepsilon_{23}k^{eu}_{1z}) + k_x^2\sin\psi\varepsilon_{11}]e^{ik^{eu}_{1z}h}}{[(k^{eu}_{1z}\sin\psi - k_y\cos\psi)^2 + k_x^2]^{1/2}[(k_x\varepsilon_{11})^2 + (\varepsilon_{22}k_y + \varepsilon_{23}k^{eu}_{1z})^2 + (\varepsilon_{32}k_y + \varepsilon_{33}k^{eu}_{1z})^2]^{1/2}}$$
$$\left.+\frac{B_{eV}[(k^{ed}_{1z}\sin\psi - k_y\cos\psi)(\varepsilon_{22}k_y + \varepsilon_{23}k^{ed}_{1z}) + k_x^2\sin\psi\varepsilon_{11}]e^{ik^{ed}_{1z}h}}{[(k^{ed}_{1z}\sin\psi - k_y\cos\psi)^2 + k_x^2]^{1/2}[(k_x\varepsilon_{11})^2 + (\varepsilon_{22}k_y + \varepsilon_{23}k^{ed}_{1z})^2 + (\varepsilon_{32}k_y + \varepsilon_{33}k^{ed}_{1z})^2]^{1/2}}\right\} \tag{5.28b}$$

The formulas (5.26–5.28) give a complete set of radiation fields for arbitrarily oriented dipoles when the dipole is embedded inside the anisotropic layer at a distance h below the top surface. As before, in spherical coordinate system, $\hat{h}$ component of the far field corresponds to $\hat{\varphi}$ component and $\hat{v}$ component of the far field corresponds to $\hat{\theta}$ component.

5.4 Physical Interpretation of Dyadic Green's Functions for Radiation Fields

The physical interpretation of dyadic Green's functions and the radiation fields derived in the previous section carries importance to understand the physical phenomenon that is occurring. This discussion is given in Sects. 5.4.1 and 5.4.2.

5.4.1 $\bar{\bar{G}}_{00}(\bar{r},\bar{r}')$: Dipole Is Placed Over the Anisotropic Layer

In the structure of the DGF, $\bar{\bar{G}}_{00}(\bar{r},\bar{r}')$ in (5.10), there are two kinds of waves. One of them is a direct wave from the source, which is represented by

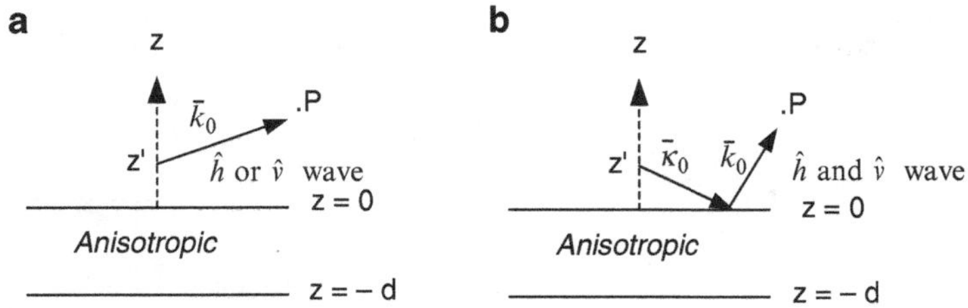

Fig. 5.4 Geometry for the physical interpretation of DGF, $\bar{\bar{G}}_{00}(\bar{r},\bar{r}')$ (**a**) Direct wave, $\hat{h}$ (or $\hat{v}$) wave. (**b**) The incident wave is $\hat{h}-$ or $\hat{v}-$ polarized. The reflected wave has both $\hat{h}-$ and $\hat{v}-$ polarized components

$\hat{h}(k_{0z})e^{-i(\bar{k}_0\cdot\bar{r}')}$ or $\hat{v}(k_{0z})e^{-i(\bar{k}_0\cdot\bar{r}')}$. Others are the reflected waves from an anisotropic surface and accompanied with appropriate reflection coefficients. There exist the cross-polarized waves, which are accompanied by the reflection coefficients, R'_{HV} or R'_{VH} and they are due to isotropic-anisotropic interface. Because when the dipole radiates a wave with either $\hat{h}-$ or $\hat{v}-$ polarization, the reflected wave from the isotropic-anisotropic boundary, will have both polarization components. For instance, R'_{HV} represents the case when the incident wave is $\hat{v}-$ polarized and the reflected wave is $\hat{h}-$ polarized. This is illustrated in Fig. 5.4. Upon observing analytical results for far fields given by (5.16–5.18), it is seen that for the horizontally oriented dipoles, each field component ($\hat{h}$ or $\hat{v}$) has terms associated with a direct wave, a wave due to image of the dipole, and a wave due to cross polarization. Hence both $\hat{h}$ component and $\hat{v}$ component of the far fields for horizontal dipole are equally excited. For the vertical oriented dipoles, mainly $\hat{v}$ component of the far fields are excited. The excitation of $\hat{h}$ component is only due to cross polarization, thus it is relatively small. This happens when there exists isotropic-anisotropic boundary. All the contributions from stratified layer are embedded in reflection coefficients.

5.4.2 $\bar{\bar{G}}_{01}(\bar{r},\bar{r}')$: *Dipole Is Embedded Inside the Anisotropic Layer*

There exist two characteristic waves in the structure of the DGF, $\bar{\bar{G}}_{01}(\bar{r},\bar{r}')$– an ordinary wave $\hat{o}$ and an extraordinary wave $\hat{e}$. Upon transmission from anisotropic medium through the isotropic region, each transmitted wave will have $\hat{h}$ and $\hat{v}$ polarized components. A_{oH}, A_{oV}, B_{oH}, B_{oV} are the transmission coefficients when the radiated wave from the source is an ordinary wave and the transmitted wave is a horizontally polarized wave or vertically polarized wave depending on the subscript associated with it. The A's represent the transmission coefficients for waves radiated upward by a source and transmitted through isotropic region without any reflection, whereas B's represent the transmission coefficients for waves radiated downward by a source and transmitted through isotropic region after reflection from

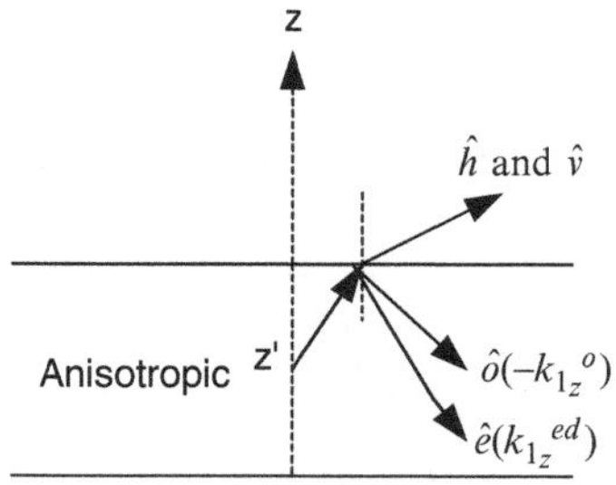

Fig. 5.5 Geometry for physical interpretation of $\bar{\bar{G}}_{01}(\bar{r},\bar{r}')$

bottom layer. By the same token, $A_{eH}, B_{eH}, A_{eV}, B_{eV}$ are the transmission coefficients when the radiated wave from the source is an extraordinary wave and the transmitted wave is a horizontally polarized wave or vertically polarized wave. Figure 5.5 illustrates this discussion.

Observing the analytical results for the far fields (5.26–5.28) reveals that the effects of orientation of the dipoles are carried by the coefficients which are in multiplication with the transmission coefficients. All the contributions from the stratified layer are embedded in the transmission coefficients.

5.5 Numerical Results

Section 5.5.1 includes discussion about analytical and numerical validation of the results given in the previous sections. In Sect. 5.5.1 some special cases have been investigated in the limiting cases and compared with existing results. In Sects. 5.5.2, 5.5.3and 5.5.4 new numerical results are presented and the effects of changing parameters are discussed in detail.

5.5.1 Special Cases

In this section, three special cases are considered to confirm the validity of the results obtained in the previous sections. Case I corresponds to the self-check of the analytical results obtained in Sect. 5.2. Case II compares the far fields in the limiting case for isotropic half space problem with the numerical results of King and Sandler [5]. Case III compares the radiation fields for the two-layered anisotropic medium when the dipole is at the interface with the numerical results of Tsalamengas and Uzunoglu [6].

5.5.1.1 Case I

The dipole is placed right at the interface between region 0 and region 1. Region 1 is assumed to be an isotropic medium having the same permitivity as region 0.

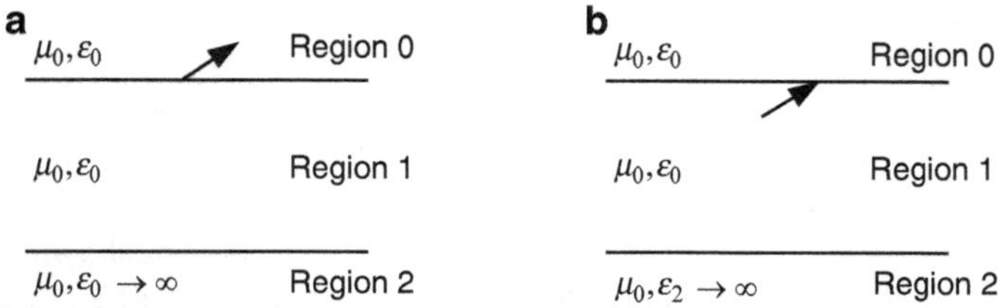

Fig. 5.6 Geometry of Case I (**a**) with DGF $\bar{\bar{G}}_{00}(\bar{r},\bar{r}')$ (**b**) with DGF $\bar{\bar{G}}_{01}(\bar{r},\bar{r}')$

Region 2 is assumed to be a ground plane. Far fields are calculated using DGFs $\bar{\bar{G}}_{00}(\bar{r},\bar{r}')$ and $\bar{\bar{G}}_{01}(\bar{r},\bar{r}')$. Figure 5.6(a) depicts the far field problem with DGF $\bar{\bar{G}}_{00}(\bar{r},\bar{r}')$, whereas Fig. 5.6(b) represents the same problem with DGF $\bar{\bar{G}}_{01}(\bar{r},\bar{r}')$.

When a vertically oriented dipole is placed at the interface (h = 0), i.e., $\hat{u}=\hat{z}$, the cross polarization coefficients $R'_{HV}\to 0, R'_{VH}\to 0$ in $G_{00}(\bar{r},\bar{r}')$. Hence, the expressions in (5.16a) and (5.16b) reduce to

$$\bar{E}_h(r)\to 0 \tag{5.29a}$$

$$\bar{E}_v(r)=\hat{v}(k_{oz})\frac{\omega\mu_0 c}{i}\frac{e^{i\bar{k}_o.\bar{r}}}{4\pi r}\frac{k_\rho}{k_o}\{1+R_{VV}\} \tag{5.29b}$$

When the same problem is solved using DGF $\bar{\bar{G}}_{01}(\bar{r},\bar{r}')$ as depicted in Fig. 5.6 (b), the transmission coefficients, off-diagonal permitivities and wave numbers reduce to

$$\begin{aligned}
\psi\to 0,\varepsilon_{22}\to 0,\ \varepsilon_{23}=\varepsilon_{32}\to 0\\
A_{eV}\to 1, A_{oH}\to 1\\
B_{eV}\to R_{VV}, B_{oH}\to R_{HH}\\
A_{eH}\to 0, B_{eH}\to 0, A_{oV}\to 0, B_{oV}\to 0\\
k_{1z}^{ed}\to -k_{1z}^{o}, k_{1z}^{eu}\to k_{1z}^{o}\\
k_1\to k_0
\end{aligned} \tag{5.30}$$

Hence, (5.26a) reduces to (5.29a) and (5.26b) reduces to (5.29b) which validates the results in one special case. When the same procedure is repeated for the horizontally oriented (x–oriented) dipole, the analytical results for the far fields using the DGFs $\bar{\bar{G}}_{00}(\bar{r},\bar{r}')$ and $\bar{\bar{G}}_{01}(\bar{r},\bar{r}')$ both reduce to the same results as follows:

$$E_h(r)=\frac{\omega\mu_0 c}{i}\frac{e^{i\bar{k}_o.\bar{r}}}{4\pi r}\left\{\frac{k_y}{k_\rho}(1+R'_{HH})\right\} \tag{5.31a}$$

$$E_v(r)=\frac{\omega\mu_0 c}{i}\frac{e^{i\bar{k}_o.\bar{r}}}{4\pi r}\left\{-\frac{k_x k_{oz}}{k_\rho k_o}(1-R'_{VV})\right\} \tag{5.31b}$$

5.5.1.2 Case II

We consider the isotropic half space problem that is solved by King & Sandler [5]. Region 0 and region 1 are assumed to be air whereas region 2 in this problem is considered to be a lossless or slightly lossy medium, i.e., dry sand or lake water. The vertical electric dipole is placed at a distance h from medium 2 as shown in Fig. 5.7. The far field radiation patterns are shown in Fig. 5.8 for two different cases, dry sand and lake water, respectively. The numerical results for radiation patterns agree well with the numerical results obtained in [5].

5.5.1.3 Case III

We consider the anisotropic two-layer problem when the dipole is located at the interface h = 0. The geometry of the problem is shown in Fig. 5.2 and Fig. 5.3. In this problem region 1 is considered to be a weakly uniaxial anisotropic medium, Polytetrafluoroethlene (PTFE, $\varepsilon_{1z} = 10.4\varepsilon_0, \varepsilon_1 = 10.7\varepsilon_0$), backed up with a perfect ground plane (region 2). The far field in region 0 is calculated using our method and shown in Fig. 5.9. The far field calculation is carried using dyadic Green's functions $\bar{\bar{G}}_{00}(\bar{r},\bar{r}')$ or $G_{01}(\bar{r},\bar{r}')$.

It is observed that for horizontally oriented dipoles, far fields obtained using $\bar{\bar{G}}_{00}(\bar{r},\bar{r}')$ or $G_{01}(\bar{r},\bar{r}')$ gave identical patterns. On the other hand, for vertically

Fig. 5.7 Vertical electric dipole over an imperfectly conducting plane. $\varepsilon_r = 2$ (dry sand), $\sigma = 0$ or $\varepsilon_r = 80$ (lake water), $\sigma = 0.004$. Dipole height = 2 m, f = 100 MHz

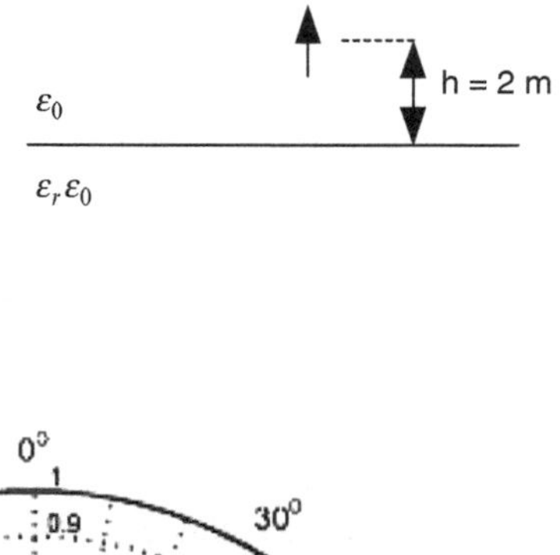

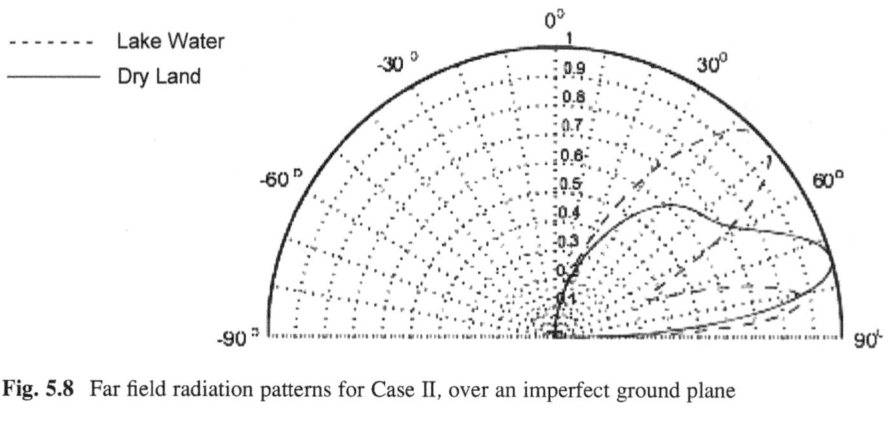

Fig. 5.8 Far field radiation patterns for Case II, over an imperfect ground plane

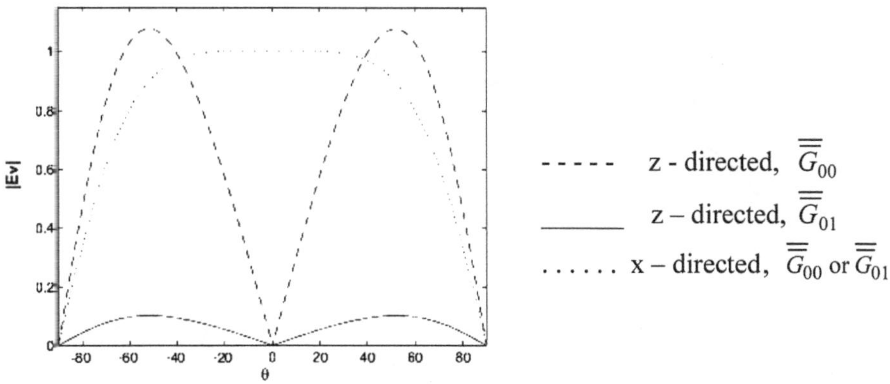

Fig. 5.9 Far field radiation patterns for *Case III,* weakly uniaxially anisotropic medium over a perfect ground plane. h $= 0$ mm, d $= 1$ mm, $\psi = 60°$, $\phi = 0°$ (180°), $\varepsilon_{1z} = 10.4\varepsilon_0$, $\varepsilon_1 = 10.7\varepsilon_0$

oriented dipoles, the far fields exhibited a difference in the amplitude, whose ratio is approximately equivalent to relative permitivity of the anisotropic medium which is weakly uniaxial and hence has a very similar characteristic to isotropic medium. The difference seen in the amplitudes of the radiation $\bar{E}$ fields for the vertically oriented dipoles is due to the continuity of the normal component of the electric flux density, $\bar{D}$ across the boundary between two media. This same problem is solved by Tsalamengas and Uzunoglu [6]. The radiation fields using $\bar{\bar{G}}_{01}(\bar{r}, \bar{r}')$ gave identical results to ones that are obtained in [6].

5.5.2 *Effect of Anisotropy*

In this section, the effect of anisotropy will be investigated on the radiation patterns for varying anisotropy when dipole is located over or embedded inside the layered anisotropic medium [7].

5.5.2.1 Dipole Is Placed Over

The geometry shown in Fig. 5.2 is used in calculation of radiation fields. We consider region 2 to be a perfect ground plane, i.e., $\varepsilon_2 \to \infty$, and the dipole is placed at the interface, h $= 0$, for practical applications. The radiation fields are obtained by varying the anisotropy parameter of the medium in region 1, $\varepsilon_{1z}/\varepsilon_1$, from 0.5 (negatively uniaxial medium) to 1.5 (positively uniaxial medium) where $\varepsilon_l = 10.7\varepsilon_0$. The medium thickness d in region 1 is 0.01λ where λ is the free space wavelength.

If the dipole orientation is in x-direction, the radiation field has a single beam when $0° < \phi \leq 90°$ $(180° < \phi \leq 270°)$ for both *h*-polarized and *v*-polarized

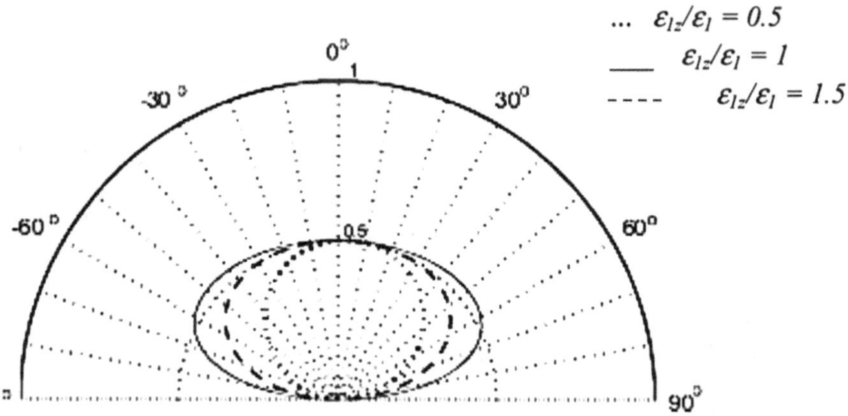

Fig. 5.10 Change in beamwidth and directivity as anisotropy is varied. |Ev| for x-directed dipole when h = 0 mm, d = 1 mm, $\psi = 60°$, $\phi = 60°(240°)$

components. The anisotropic effect on the radiation fields when the observation point is on the $\phi = 90°$ (270°) plane is minimal for the *h*-polarized component of the radiation field. On the other hand, this effect is significant for the *v*-polarized component for the same dipole orientation. As the observation point moves to the $\phi = 0°(180°)$ plane, this effect is reversed for the *h*-polarized and *v*-polarized components. For the *v*-polarized component, the radiation intensity is strongest when the medium is positively uniaxial ($\varepsilon_{1z}/\varepsilon_1 = 1.5$) and the weakest when the medium is negatively uniaxial ($\varepsilon_{1z}/\varepsilon_1 = 0.5$). Figure 5.10 shows the radiation patterns for the v-polarized wave when the dipole is oriented in x-direction and $\phi = 60°$ for $0° \le \theta \le 90°$ or $\phi = 240°$ for $-90° \le \theta \le 0°$. As the anisotropy becomes more positively uniaxial, the beamwidth gets broader with the maximum beam at $\theta = 48.2°$ and the directivity decreases for the *v*-polarized waves. As a consequence of this effect, the beamwidth and the directivity of the antenna can be controlled with the anisotropy parameter. When the medium is positively uniaxial, i.e., $\varepsilon_{1z}/\varepsilon_1 = 1.5$, the main beam shifts towards $\theta = 90°$ as the observation plane moves towards $\phi = 0°(180°)$ plane. This is similar to a change of radiation pattern from an endfire antenna to a broadside antenna. This effect is reversed for the *h*-polarized waves. It is important to note that when $\phi = 0°(180°)$, the *v*-polarized component of the radiation field will be dominant. If the observation plane is shifted to $\phi = 90°(270°)$ plane, the *h*-polarized component will be dominant. When $\phi = 45°(225°)$, both components will contribute equally to the radiation field.

When the dipole is oriented in z-direction with $\phi = 60°(240°)$, both *v*-polarized and *h*-polarized components of the radiation field has a double beam pattern. The radiation field intensity changes in the same way for the *h*-polarized and *v*-polarized components which was described for x-oriented dipoles when $0° \le \phi \le 90°$ ($180° \le \phi \le 270°$). Since the radiation field has a double beam pattern consistently, this corresponds to change in the lobe height of the radiation pattern. The *h*-polarized component of the radiation field is zero when region 1 is isotropic since there is no cross-polarized component.

5.5.2.2 Dipole Is Embedded Inside

Here, the geometry of Fig. 5.3 is considered in calculation of radiation fields. The dipole is embedded in the middle of the anisotropic medium, h = 0.05λ. The thickness of the medium, d, and the anisotropy parameter are kept the same. The anisotropy parameter is varied in the same way as in Sect. 5.5.2.1 to calculate the radiation fields. The *h*-polarized component of the z-oriented dipole has a single beam when $0° \leq \phi \leq 90°$ ($180° \leq \phi \leq 270°$). The numerical results show that the radiation intensity and the maximum beam change by varying the anisotropy parameter. This corresponds to main beam movement for the radiation pattern of an antenna with a single beam. The *v*-polarized component of the radiation field exhibits very interesting case. If we consider region 1 as an isotropic medium, the radiation pattern has a double beam when $0° \leq \phi \leq 90°$ ($180° \leq \phi \leq 270°$). When the medium is negatively uniaxial or positively uniaxial, it has a single beam pattern on the same observation plane except when $\phi = 90°(270°)$. This corresponds to change in the radiation pattern. Figure 5.11 shows the numerical results for v-polarized component of the radiation field when $\phi = 60°(240°)$.

In this figure, when the medium in region 1 is isotropic, the radiation pattern has double beam which has null at $\theta = 0°$ and maximum radiation at $\theta = \pm 55.5°$. When the medium is negatively uniaxial ($\varepsilon_{1z}/\varepsilon_1 = 0.5$) or positively uniaxial ($\varepsilon_{1z}/\varepsilon_1 = 1.5$), this double beam pattern switches to single beam pattern with higher radiation intensity for the negatively uniaxial medium in comparison to the positively uniaxial medium and isotropic medium, respectively. The maximum radiation occurs at $\theta = 19.1°$ for negatively uniaxial case whereas this point changes to $\theta = 51.8°$ for positively uniaxial case.

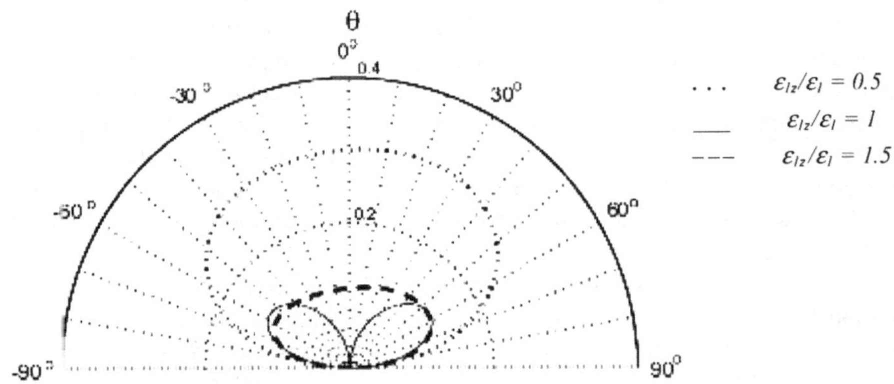

Fig. 5.11 Change in pattern shape as anisotropy is varied. |Ev| for z-directed dipole when h = 0.05λ, d = 0.1λ, ψ = 60°, $\phi = 60°(240°)$

The change in the radiation fields for the x-oriented dipole by varying the anisotropy parameter is similar when it is placed at the interface.

5.5.3 *Effect of Layer Thickness*

In this section, the effect of layer thickness will be investigated on the radiation patterns for different thickness levels when dipole is located over or embedded inside the layered anisotropic medium.

5.5.3.1 Dipole Is Over

The radiation fields are obtained when region 1 is a negatively uniaxial medium, Epsilam 10 ($\varepsilon_{1z} = 10.2\varepsilon_0, \varepsilon_1 = 13\varepsilon_0$) or a positively uniaxial medium, Sapphire ($\varepsilon_{1z} = 11.6\varepsilon_0, \varepsilon_1 = 9.4\varepsilon_0$). The dipole is assumed to be at the interface, h = 0 mm, with orientation in x-direction or z-direction. The layer thickness of the uniaxial medium is varied from 0.1λ to 1λ.

If the dipole is oriented in the x-direction and the medium in region 1 is Epsilam 10 with thickness 1λ, the radiation pattern for *h*-polarized component in the $\phi = 90°(270°)$ plane has one main lobe with max radiation at $\theta = 0.91°$ and two symmetrical minor lobes. If the observation point is taken such that $0° < \phi < 90°$ $(180° < \phi < 270°)$, the radiation field has a single beam pattern. When $\phi = 0°(180°)$ plane, although the intensity of the radiation field is smaller, the radiation pattern becomes symmetrical with two lobes having maximum radiation at $\theta = \pm 50°$. If layer thickness of the anisotropic medium is changed to 0.1λ, the radiation pattern has as a single beam when $0° < \phi < 90°$ $(180° < \phi < 270°)$ with max radiation at $\theta = 0.909°$ and switches to a symmetrical, double beam pattern with maximum radiation at $\theta = \pm 35.5$ on $\phi = 0°(180°)$ plane. Hence, the effect of varying the layer thickness can be used to control the main beam movement of the radiation pattern for the *h*-polarized component of the x-oriented dipole when $0° < \phi < 90°$ $(180° < \phi < 270°)$. This effect is shown in Fig. 5.12. In this figure, the maximum radiation occurs at $\theta = -11.8°$ or $\theta = -0.91°$ for layer thickness 1λ or 0.1λ respectively, when $\phi = 45°(225°)$. The radiation patterns for the *v*-polarized component of the x-oriented dipole are almost symmetrical (with respect to ϕ) to those for the *h*-polarized component of the radiation field for x-directed dipole.

When the orientation of the dipole is changed to z-direction and the medium in region 1 is Epsilam 10, the *h*-polarized component which has only a cross polarization term, has a double beam radiation pattern when $0° \leq \phi \leq 90°$ $(180° \leq \phi \leq 270°)$ for 1λ or 0.1λ with different lobe heights. The *v*-polarized component of the dipole has its maximum radiation when $\phi = 0°(180°)$ and it has double beam pattern everywhere except $\phi = 90°(270°)$ and layer thickness is 1λ. At this

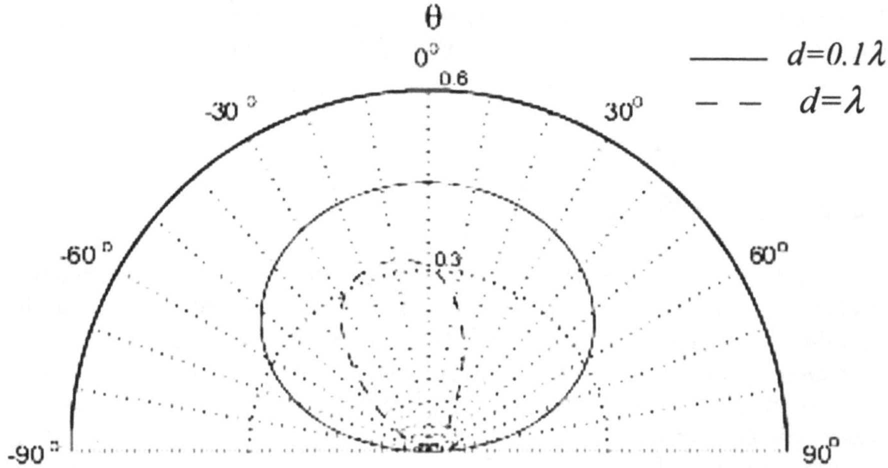

Fig. 5.12 Main beam movement as a result of change in thickness of the material. |Eh| for x-directed dipole over the layer. h = 0 mm, $\psi = 60°$, $\phi = 45°(225°)$, $\varepsilon_{1z} = 10.3\varepsilon_0$, $\varepsilon_1 = 13\varepsilon_0$

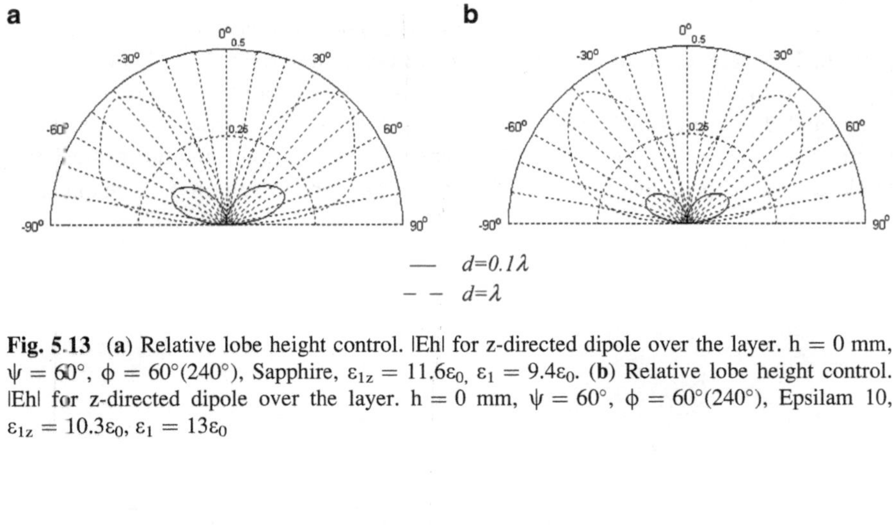

Fig. 5.13 (**a**) Relative lobe height control. |Eh| for z-directed dipole over the layer. h = 0 mm, $\psi = 60°$, $\phi = 60°(240°)$, Sapphire, $\varepsilon_{1z} = 11.6\varepsilon_0$, $\varepsilon_1 = 9.4\varepsilon_0$. (**b**) Relative lobe height control. |Eh| for z-directed dipole over the layer. h = 0 mm, $\psi = 60°$, $\phi = 60°(240°)$, Epsilam 10, $\varepsilon_{1z} = 10.3\varepsilon_0$, $\varepsilon_1 = 13\varepsilon_0$

observation point with layer thickness 1λ, it has additional two symmetrical minor lobes. So, when the observation point is $0° \leq \phi < 90°$ ($180° \leq \phi < 270°$), the layer thickness can be used to control the relative lobe height of the radiation pattern. When the medium in region 1 is replaced with Sapphire, the radiation patterns change in the similar way with different intensities and maximum radiation points.

The relative lobe height control by changing the layer thickness is shown in Fig. 5.13(a) when the medium 1 is Sapphire and in Fig. 5.13(b) when the medium 1 is Epsilam 10, for the *h*-polarized component of the radiation field for the z-directed dipole. The maximum radiation occurs at $\theta = 42.7°$ with magnitude of 0.49 when d = 1λ for Sapphire. This maximum beam shifts to at $\theta = 60.9°$ with magnitude

0.18 when d = 0.1λ. In the case of Epsilam 10, the maximum radiation occurs at θ = 37.3° with magnitude of 0.42 and θ = 59.1° with magnitude of 0.13 when d = 1λ and d = 0.1λ, respectively.

5.5.3.2 Dipole Is Embedded Inside

In this section, dipole is located in the middle of the uniaxial anisotropic medium, Epsilam 10 or Sapphire, in region 1. The radiation fields are calculated when the layer thickness of this medium is varied. When the dipole is oriented in z-direction and region 1 is Epsilam 10 with layer thickness 0.1λ or 1λ, the radiation field has a single beam pattern for *h*-polarized component with a reduced radiation field intensitiy in comparison to Sect. 5.5.3.1 for $0° \leq \phi \leq 90°$ ($180° \leq \phi \leq 270°$). It is worth to mention that the directivity of the antenna improves on $\phi = 0°(180°)$ plane when the anisotropic medium has a layer thickness 1λ. The v-polarized component of the radiation field for the same dipole orientation has a single beam pattern when $0° < \phi \leq 90°$ ($180° < \phi \leq 270°$) with layer thickness 0.1λ or 1λ. As we approach to the $\phi = 0°(180°)$ plane the intensity of the radiation field decreases. When $\phi = 0°(180°)$, the radiation field switches to a double beam pattern. This behaviour differs for the same orientation in Sect. 5.5.3.1. Figure 5.14 represents the main beam movement for the *v*-polarized component of the z-directed dipole in Epsilam 10 when $\phi = 60°$ (240°). The maximum radiation happens at $\theta = \pm 39.1°$ or $\theta = \pm 53.6°$ when the anisotropic medium has a layer thickness 1λ or 0.1λ, respectively. When the medium in region 1 is changed to

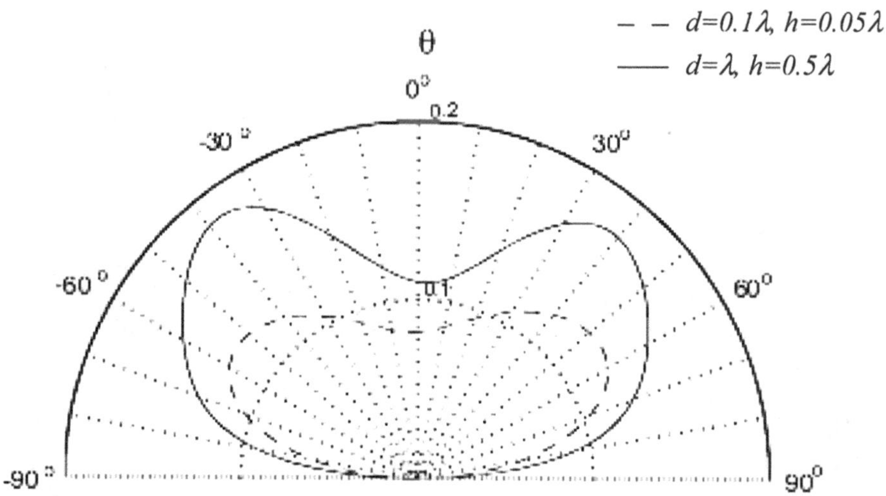

Fig. 5.14 Change in relative lobe height. |Ev| for z-directed dipole inside the layer. $\psi = 60°$, $\phi = 60°(240°)$, Epsilam 10, $\varepsilon_{1z} = 10.3\varepsilon_0$, $\varepsilon_1 = 13\varepsilon_0$

Sapphire, the observations made in Sect. 5.5.3.1 hold for the change in the radiation fields for both h- and v-polarized components for this orientation.

The radiation fields of h-polarized component of the radiation field for the x-oriented dipole when the anisotropic medium is Epsilam 10, have single lobe except when the observation point is on the $\phi = 0°(180°)$ plane. On this plane the pattern has a double beam pattern for the layer thickness 1λ or 0.1λ. For h-polarized component, the change in the radiation fields on different observation planes as a response to varying layer thickness is similar to the changes in Sect. 5.5.3.1 for the same dipole orientation. The v-polarized component of the radiation field for this dipole orientation has a single beam on all observation planes unlike the v-polarized component of the same orientation in the previous section. The radiation field intensity increases as observation point moves to $\phi = 0°(180°)$ plane. When the medium in region 1 is Sapphire, similar observations in Sect. 5.5.3.1 hold.

5.5.4 Effect of Dipole Location

In this section, the radiation fields are obtained by varying the dipole location between 0.01λ, which is a closer point to the region 1 – region 0 boundary, to 0.09λ, which is a closer point to the region 1 – region 2 boundary, inside a uniaxial medium, Epsilam 10 or Sapphire. The thickness of the medium is considered to be 0.1λ.

When the orientation of the dipole is in z-direction and the anisotropic medium is Epsilam 10, the h-polarized component of the radiation field has a single beam on the observation plane $0° \leq \phi \leq 90°$ $(180° \leq \phi \leq 270°)$ for two different dipole location. The radiation intensity is stronger when the dipole location is closer to the region 1- region 0 boundary, $h = 0.01\lambda$. The maximum radiation occurs at $\theta = 11.8°$ when the observation point is on y-z plane $(\phi = 90°)$, whereas it happens at $\theta = 0.91°$ if the observation point is on x-z plane $(\phi = 0°)$. The maximum radiation is observed at $\theta = -39.1°$ on both $\phi = 270°$ plane and $\phi = 180°$ plane if the dipole location is at $h = 0.09\lambda$. The v-polarized component of the radiation field has a single beam when the dipole location is closer to region 1- region 0 boundary between the observation planes $\phi = 90°(270°)$ and $\phi = 0°$ $(180°)$, whereas the radiation pattern has a double beam symmetrical pattern when the dipole location is closer to region 1-region 2 boundary. This change in the radiation pattern is shown in Fig. 5.15 when $\phi = 60°$ $(240°)$.

If the orientation of the dipole is changed to x-direction for the same anisotropic medium in region 1, the numerical results showed that both v- and h-polarized components have single beams between observation planes $\phi = 90°(270°)$ and $\phi = 0°(180°)$. Although the radiation intensity is small, h-polarized component exhibits a double beam pattern on the $\phi = 0°(180°)$ plane.

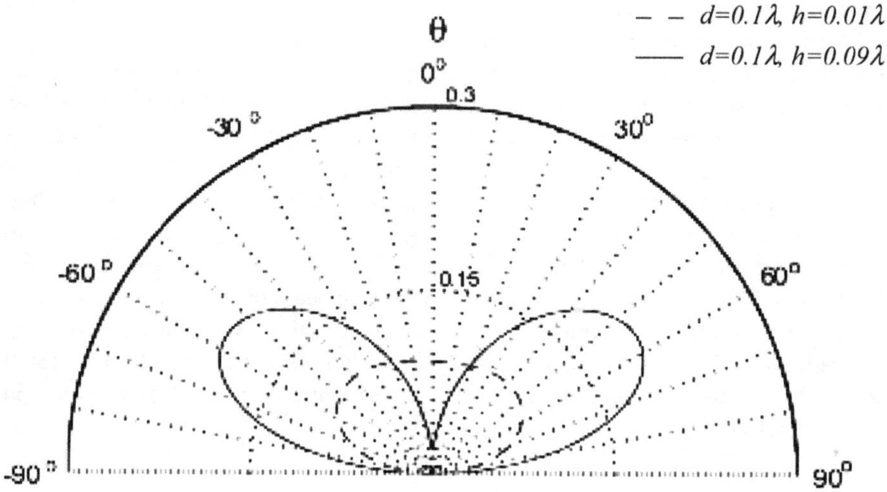

Fig. 5.15 Change from single to double beam. |Ev| for z-directed dipole inside the layer. $\psi = 60°$, $\phi = 60°(240°)$, $\varepsilon_{1z} = 10.3\varepsilon_0$, $\varepsilon_1 = 13\varepsilon_0$

Appendix

The wave vectors used in the DGF $\bar{\bar{G}}_{00}(\bar{r},\bar{r}')$ and $\bar{\bar{G}}_{01}(\bar{r},\bar{r}')$ are:

$$\bar{k}_0 = \bar{k}_\rho + k_{0z}\hat{z} \tag{5.32}$$

$$\bar{\kappa}_0 = \bar{k}_\rho - k_{0z}\hat{z} \tag{5.33}$$

$$\bar{k}_1^o = \bar{k}_\rho + k_{1z}^o\hat{z} \tag{5.34}$$

$$\bar{\kappa}_1^o = \bar{k}_\rho - k_{1z}^o\hat{z} \tag{5.35}$$

$$\bar{k}_1^e = \bar{k}_\rho + k_{1z}^{eu}\hat{z} \tag{5.36}$$

$$\bar{\kappa}_1^e = \bar{k}_\rho + k_{1z}^{ed}\hat{z} \tag{5.37}$$

$$\bar{k}_\rho = \hat{x}k_x + \hat{y}k_y \tag{5.38}$$

The unit vectors for the horizontally polarized and vertically polarized waves are:

$$\hat{h}(k_{0z}) = \frac{1}{k_\rho}\hat{z} \times \bar{k}_0 \tag{5.39a}$$

$$\hat{h}(-k_{0z}) = \frac{1}{k_\rho}\hat{z} \times \bar{\kappa}_0 \tag{5.39b}$$

$$\hat{v}(k_{0z}) = \frac{1}{k_0}\hat{h}(k_{0z}) \times \bar{k}_0 \tag{5.40a}$$

$$\hat{v}(-k_{0z}) = \frac{1}{k_0}\hat{h}(-k_{0z}) \times \bar{\kappa}_0 \tag{5.40b}$$

The unit vectors for an ordinary and an extraordinary waves are:

$$\hat{o}(k_z^o) = \frac{\hat{z}' \times \bar{k}^o}{\left|\hat{z}' \times \bar{k}^o\right|} \tag{5.41a}$$

$$\hat{o}(-k_z^o) = \frac{\hat{z}' \times \bar{\kappa}^o}{\left|\hat{z}' \times \bar{\kappa}^o\right|} \tag{5.41b}$$

$$\hat{e}(k_z^{eu}) = \frac{\hat{o}(k_z^{eu}) \times \bar{k}_u}{\left|\bar{k}_u\right|} \tag{5.42a}$$

$$\hat{e}(k_z^{ed}) = \frac{\hat{o}(k_z^{ed}) \times \bar{\kappa}_u}{\left|\bar{\kappa}_u\right|} \tag{5.42b}$$

where

$$\bar{k}^o = \bar{k}_\rho + k_z^o \hat{z} \tag{5.43a}$$

$$\bar{\kappa}^o = \bar{k}_\rho - k_z^o \hat{z} \tag{5.43b}$$

$$\bar{k}^e = \bar{k}_\rho + k_z^{eu} \hat{z} \tag{5.44a}$$

$$\bar{\kappa}^e = \bar{k}_\rho + k_z^{ed} \hat{z} \tag{5.44b}$$

$$\bar{k}_u = \bar{\bar{\varepsilon}}.\bar{k}^e \tag{5.45}$$

$$\bar{\kappa}_u = \bar{\bar{\varepsilon}}.\bar{\kappa}^e \tag{5.46}$$

$$\hat{z}' = \hat{y}\sin\psi + \hat{z}\cos\psi \tag{5.47}$$

The wave numbers are defined as

$$k_{iz} = \sqrt{k_i^2 - k_\rho^2}, \quad i = 0, 2 \tag{5.48}$$

$$k_{1z}^{o} = \sqrt{k_1^2 - k_\rho^2} \tag{5.49}$$

$$\begin{bmatrix} k_{1z}^{eu} \\ k_{1z}^{ed} \end{bmatrix} = -\frac{\varepsilon_{23}}{\varepsilon_{33}} k_y \pm \frac{1}{\varepsilon_{33}} \left[k_1^2 \varepsilon_{1z} \varepsilon_{33} - \varepsilon_1 \varepsilon_{33} k_x^2 - \varepsilon_1 \varepsilon_{1z} k_y^2 \right]^{1/2} \tag{5.50}$$

$$k_i = \omega \sqrt{\mu \varepsilon_i}, \quad i = 0, 1, 2 \tag{5.51}$$

The expressions for two-layer coefficients R_{HH}, R_{HV}, R_{VV}, R_{VH}, A_{Ho}, A_{Vo}, etc. are:

$$\begin{aligned} R_{\beta\alpha} = {} & R_{01\beta\alpha} + X_{\beta o}(L_1 o_o + M_2 e_o) X_{o\alpha} + X_{\beta o}(L_1 o_e + M_2 e_e) X_{e\alpha} \\ & + X_{\beta e}(L_2 o_o + M_1 e_o) X_{o\alpha} + X_{\beta e}(L_2 o_e + M_1 e_e) X_{e\alpha} \end{aligned} \tag{5.52}$$

$$A_{\beta o} = X_{\beta o} L_1 + X_{\beta e} L_2 \tag{5.53}$$

$$A_{\beta e} = X_{\beta o} M_2 + X_{\beta e} M_1 \tag{5.54}$$

$$B_{\beta\gamma} = X_{\beta o}\left(L_1 o_\gamma + M_2 e_\gamma\right) + X_{\beta e}\left(L_2 o_\gamma + M_1 e_\gamma\right) \tag{5.55}$$

$$X_{\beta\alpha} = X_{\beta o}(L_1 t_{o\alpha} + M_2 t_{e\alpha}) + X_{\beta e}(L_2 t_{o\alpha} + M_1 t_{e\alpha}) \tag{5.56}$$

where

$$\beta, \alpha = H \text{ or } V$$

$$\gamma = o \text{ or } e$$

$$L_1 = \frac{1 - S_{ee}}{D} \tag{5.57}$$

$$L_2 = \frac{S_{eo}}{D} \tag{5.58}$$

$$M_1 = \frac{1 - S_{oo}}{D} \tag{5.59}$$

$$M_2 = \frac{S_{oe}}{D} \tag{5.60}$$

$$S_{oo} = o_o R_{oo} + o_e R_{eo} \tag{5.61}$$

$$S_{oe} = o_o R_{oe} + o_e R_{ee} \tag{5.62}$$

$$S_{eo} = e_o R_{oo} + e_e R_{eo} \tag{5.63}$$

$$S_{ee} = e_o R_{oe} + e_e R_{ee} \tag{5.64}$$

$$D = (1 - S_{oo})(1 - S_{ee}) - S_{oe} S_{eo} \tag{5.65}$$

$$X_{He} = -\frac{U_d}{D_e}\frac{2k_{oz}}{k_\rho} k_x \left(k_\rho^2 + k_{oz} k_{1z}^o\right) \sin\psi \tag{5.66}$$

$$\begin{aligned} X_{Ho} = &-\frac{g_d}{D_e}\frac{2}{\left(k_{oz} + k_{1z}^o\right)}\frac{k_{oz}}{k_\rho} \\ &\times \left\{k_\rho^2 \left(k_1^2 k_{oz} - k_0^2 k_{1z}^{ed}\right)\cos\psi + k_y \left(k_0^2 k_{1z}^{o\,2} - k_1^2 k_{oz} k_{1z}^{ed}\right)\sin\psi\right\} \end{aligned} \tag{5.67}$$

$$R_{01HH} = -1 + \frac{X_{Ho}}{g_d}\frac{1}{k_\rho}\left(k_\rho^2 \cos\psi + k_y k_{1z}^o \sin\psi\right) - \frac{k_1^2}{k_\rho} k_x \sin\psi \frac{X_{He}}{U_d} \tag{5.68}$$

$$R_{C1HV} = -\frac{X_{Ho}}{g_d}\frac{k_0 k_x}{k_\rho k_{oz}} k_{1z}^o \sin\psi + \frac{X_{He}}{U_d}\frac{k_0}{k_\rho k_{oz}}\left(k_\rho^2 k_{1z}^{ed}\cos\psi - k_{1z}^{o\,2} k_y \sin\psi\right) \tag{5.69}$$

$$\begin{aligned} D_e = &\cos^2\psi k_\rho^2 \left(k_1^2 k_{oz} - k_0^2 k_{1z}^{ed}\right) + \sin^2\psi\left\{k_1^2\left(k_{oz} - k_{1z}^{ed}\right)\left(k_x^2 + k_{oz}k_{1z}^o\right) + k_y^2 k_{1z}^o \left(k_1^2 - k_0^2\right)\right\} \\ &+ \cos\psi\sin\psi k_y \left(k_{1z}^o + k_{oz}\right)\left(k_{1z}^o - k_{1z}^{ed}\right)\left(k_\rho^2 + k_{1z}^o k_{oz}\right) \end{aligned} \tag{5.70}$$

$$U_d = \left\{\frac{\varepsilon_1}{\varepsilon_1 - \varepsilon_{1z}}\left(k_x^2 + k_y^2 + k_{1z}^{ed^2} - \omega^2\mu\varepsilon_1\right)\left[k_x^2 + k_y^2 + k_{1z}^{ed^2} - \omega^2\mu(\varepsilon_1 + \varepsilon_{1z})\right]\right\}^{1/2} \tag{5.71}$$

$$g_d = \left\{k_x^2 + \left(k_y \cos\psi + k_{1z}^o \sin\psi\right)^2\right\}^{1/2} \tag{5.72}$$

$$X_{Ve} = \frac{U_d}{D_e}\frac{2k_0}{k_\rho}\left(k_{oz} k_\rho^2 \cos\psi + k_y k_{oz} k_{1z}^o \sin\psi\right) \tag{5.73}$$

$$X_{Vo} = \frac{g_d}{D_e}\frac{2k_0 k_{oz} k_1^2 k_x}{k_\rho}\frac{k_{oz} - k_{1z}^{ed}}{k_{oz} + k_{1z}^o}\sin\psi \tag{5.74}$$

$$R_{0\_VV} = 1 - \frac{X_{Vo}}{g_d}\frac{k_0 k_x}{k_\rho k_{oz}} k_{1z}^o \sin\psi + \frac{X_{Ve}}{U_d}\frac{k_0}{k_\rho k_{oz}}\left(k_\rho^2 k_{1z}^{ed}\cos\psi - k_{1z}^{o^2} k_y \sin\psi\right) \tag{5.75}$$

$$R_{01VH} = \frac{X_{Vo}}{g_d}\frac{1}{k_\rho}\left(k_\rho^2\cos\psi + k_y k_{1z}^o \sin\psi\right) - \frac{k_1^2}{k_\rho} k_x \sin\psi \frac{X_{Ve}}{U_d} \tag{5.76}$$

The coefficients $o_o, e_o, \mathrm{o}_e, e_e$ and $t_{oH}, \mathrm{t}_{eH}, \mathrm{t}_{oV}, \mathrm{t}_{eV}$ are:

$$\begin{bmatrix} o_o & e_o \\ o_e & e_e \end{bmatrix} = \begin{bmatrix} e^{ik_{1z}^o d} R_{12oo} e^{ik_{1z}^o d} & e^{-ik_{1z}^{ed} d} R_{12eo} e^{ik_{1z}^o d} \\ e^{ik_{1z}^o} R_{12oe} e^{ik_{1z}^{eu} d} & e^{-ik_{1z}^{ed} d} R_{12ee} e^{ik_{1z}^{eu} d} \end{bmatrix} \tag{5.77}$$

$$\begin{bmatrix} t_{oH} & t_{eH} \\ t_{oV} & t_{eV} \end{bmatrix} = \begin{bmatrix} e^{ik_{1z}^o d} X_{12oH} e^{-ik_{2z} d} & e^{-ik_{1z}^{ed} d} X_{12eH} e^{-ik_{2z} d} \\ e^{ik_{1z}^o} X_{12oV} e^{-ik_{2z} d} & e^{-ik_{1z}^{ed} d} X_{12eV} e^{-ik_{2z} d} \end{bmatrix} \tag{5.78}$$

where

$$R_{12oo} = \frac{g_u}{g_d} \frac{G_e}{F_e} \frac{k_{1z}^o - k_{2z}}{k_{1z}^o + k_{2z}} \tag{5.79}$$

$$R_{12oe} = \frac{U_u}{g_d} \frac{1}{F_e} 2k_x k_{1z}^o \left(k_{2z} - k_{1z}^o\right)\left(k_y \sin\psi + k_{2z}\cos\psi\right)\sin\psi \tag{5.80}$$

$$\begin{aligned} X_{12oH} =& \frac{1}{g_d k_\rho}\left(k_\rho^2 \cos\psi + k_y k_{1z}^o \sin\psi\right) + \frac{R_{12oo}}{g_u k_\rho}\left(k_\rho^2\cos\psi - k_y k_{1z}^o \sin\psi\right) \\ & - \frac{R_{12oe}}{U_u k_\rho} k_1^2 k_x \sin\psi \end{aligned} \tag{5.81}$$

$$\begin{aligned} X_{12oV} =& \frac{k_2}{k_{2z} k_\rho}\left\{\frac{1}{g_d} k_x k_{1z}^o \sin\psi - \frac{R_{12oo}}{g_u} k_x k_{1z}^o \right. \\ & \left. \times \sin\psi + \frac{R_{12oe}}{U_u}\left(k_y k_{1z}^{o^2} \sin\psi - k_\rho^2 k_{1z}^{eu}\cos\psi\right)\right\} \end{aligned} \tag{5.82}$$

$$\begin{aligned} F_e =& \cos^2\psi k_\rho^2\left(k_1^2 k_{2z} + k_2^2 k_{1z}^{eu}\right) + \sin^2\psi\left\{k_1^2\left(k_{2z} + k_{1z}^{eu}\right)\left(k_x^2 + k_{2z}k_{1z}^o\right) + k_y^2 k_{1z}^o\left(k_1^2 - k_2^2\right)\right\} \\ & - \cos\psi\sin\psi k_y\left(k_{1z}^o + k_{2z}\right)\left(k_{1z}^o + k_{1z}^{eu}\right)\left(k_\rho^2 + k_{1z}^o k_{2z}\right) \end{aligned} \tag{5.83}$$

$$\begin{aligned} G_e =& \cos^2\psi k_\rho^2\left(k_1^2 k_{2z} + k_2^2 k_{1z}^{eu}\right) + \sin^2\psi\left\{k_1^2\left(k_{2z} + k_{1z}^{eu}\right)\left(k_x^2 - k_{2z}k_{1z}^o\right) - k_y^2 k_{1z}^o\left(k_1^2 - k_2^2\right)\right\} \\ & + \cos\psi\sin\psi k_y\left(k_{2z} - k_{1z}^o\right)\left(k_{1z}^o - k_{1z}^{eu}\right)\left(k_\rho^2 - k_{1z}^o k_{2z}\right) \end{aligned} \tag{5.84}$$

$$U_u = \left\{\frac{\varepsilon_1}{\varepsilon_1 - \varepsilon_{1z}}\left(k_x^2 + k_y^2 + k_{1z}^{ed^2} - \omega^2\mu\varepsilon_1\right)\left[k_x^2 + k_y^2 + k_{1z}^{eu^2} - \omega^2\mu(\varepsilon_1 + \varepsilon_{1z})\right]\right\}^{1/2} \tag{5.85}$$

$$g_u = \left\{k_x^2 + \left(k_y\cos\psi - k_{1z}^o \sin\psi\right)^2\right\}^{1/2} \tag{5.86}$$

$$R_{12ee} = -\frac{U_u}{U_d}\frac{H_e}{F_e} \tag{5.87}$$

$$R_{12eo} = \frac{g_u}{U_d}\frac{1}{F_e}k_1^2 k_x\left(k_{1z}^{eu} - k_{1z}^{ed}\right)\left(k_{1z}^o - k_{2z}\right)\left(k_y \sin\psi - k_{2z}\cos\psi\right)\sin\psi \tag{5.88}$$

$$X_{12eV} = \frac{k_2}{k_{2z}k_\rho}\left\{\frac{1}{U_d}\left(k_{1z}^{o^2}k_y \sin\psi - k_\rho^2 k_{1z}^{ed}\cos\psi\right) - \frac{R_{12eo}}{g_u}k_x k_{1z}^o \sin\psi \right.$$
$$\left. + \frac{R_{12ee}}{U_u}\left(k_y k_{1z}^{o^2}\sin\psi - k_\rho^2 k_{1z}^{eu}\cos\psi\right)\right\} \tag{5.89}$$

$$X_{12eH} = \frac{1}{k_\rho}\left\{-\frac{k_1^2}{U_d}k_x \sin\psi + \frac{R_{12eo}}{g_u}\times\left(k_\rho^2\cos\psi - k_y k_{1z}^o \sin\psi\right) - \frac{R_{12ee}}{U_u}k_1^2 k_x \sin\psi\right\} \tag{5.90}$$

$$H_e = \cos^2\psi k_\rho^2\left(k_1^2 k_{2z} + k_2^2 k_{1z}^{ed}\right) + \sin^2\psi\left\{k_1^2\left(k_{2z} + k_{1z}^{ed}\right)\left(k_x^2 + k_{2z}k_{1z}^o\right) + k_y^2 k_{1z}^o\left(k_1^2 - k_2^2\right)\right\}$$
$$- \cos\psi \sin\psi k_y\left(k_{1z}^o + k_{2z}\right)\left(k_{1z}^o + k_{1z}^{ed}\right)\left(k_\rho^2 + k_{1z}^o k_{2z}\right) \tag{5.91}$$

$$R_{oo} = \frac{g_d}{g_u}\frac{E_e}{D_e}\frac{k_{1z}^o - k_{oz}}{k_{1z}^o + k_{oz}} \tag{5.92}$$

$$R_{oe} = \frac{U_d}{g_u}\frac{1}{D_e}2k_x k_{1z}^o\left(k_{1z}^o - k_{oz}\right)\left(k_y \sin\psi - k_{oz}\cos\psi\right)\sin\psi \tag{5.93}$$

$$X_{oH} = \frac{1}{g_u k_\rho}\left(k_\rho^2\cos\psi - k_y k_{1z}^o \sin\psi\right) + \frac{R_{oo}}{g_d k_\rho}\left(k_\rho^2\cos\psi + k_y k_{1z}^o\sin\psi\right)$$
$$- \frac{R_{oe}}{U_d k_\rho}k_1^2 k_x \sin\psi \tag{5.94}$$

$$X_{oV} = \frac{k_0}{k_{oz}k_\rho}\left\{\frac{1}{g_u}k_x k_{1z}^o \sin\psi - \frac{R_{oo}}{g_d}k_x k_{1z}^o \right.$$
$$\left. \times \sin\psi - \frac{R_{oe}}{U_d}\left(k_y k_{1z}^{o^2}\sin\psi - k_\rho^2 k_{1z}^{eu}\cos\psi\right)\right\} \tag{5.95}$$

$$E_e = \cos^2\psi k_\rho^2\left(k_1^2 k_{oz} - k_0^2 k_{1z}^{ed}\right) + \sin^2\psi\left\{k_1^2\left(k_{oz} - k_{1z}^{ed}\right)\left(k_x^2 - k_{oz}k_{1z}^o\right) - k_y^2 k_{1z}^o\left(k_1^2 - k_0^2\right)\right\}$$
$$+ \cos\psi \sin\psi k_y\left(k_{1z}^o - k_{oz}\right)\left(k_{1z}^o + k_{1z}^{ed}\right)\left(k_\rho^2 - k_{1z}^o k_{oz}\right) \tag{5.96}$$

$$R_{ee} = -\frac{U_d}{Uu}\frac{I_e}{D_e} \tag{5.97}$$

$$R_{eo} = \frac{g_d}{Uu}\frac{1}{D_e}k_1^2 k_x\left(k_{1z}^{ed} - k_{1z}^{eu}\right)\left(k_{1z}^o - k_{oz}\right)\left(k_y \sin\psi + k_{oz}\cos\psi\right)\sin\psi \tag{5.98}$$

$$X_{eV} = -\frac{k_0}{k_{oz}k_\rho}\left\{\frac{1}{U_u}\left(k_{1z}^{o^2}k_y \sin\psi - k_\rho^2 k_{1z}^{eu}\cos\psi\right) + \frac{R_{eo}}{g_d}k_x k_{1z}^o \sin\psi \right.$$
$$\left. + \frac{R_{ee}}{U_d}\left(k_y k_{1z}^{o^2}\sin\psi - k_\rho^2 k_{1z}^{ed}\cos\psi\right)\right\} \tag{5.99}$$

$$X_{eH} = \frac{1}{k_\rho}$$
$$\times\left\{-\frac{k_1^2}{U_u}k_x \sin\psi + \frac{R_{eo}}{g_d}\left(k_\rho^2\cos\psi + k_y k_{1z}^o \sin\psi\right) - \frac{R_{ee}}{U_d}k_1^2 k_x \sin\psi\right\} \tag{5.100}$$

$$I_e = \cos^2\psi k_\rho^2\left(k_1^2 k_{oz} - k_0^2 k_{1z}^{eu}\right) + \sin^2\psi\left\{k_1^2\left(k_{oz} - k_{1z}^{eu}\right)\left(k_x^2 + k_{oz}k_{1z}^o\right) + k_y^2 k_{1z}^o\left(k_1^2 - k_0^2\right)\right\}$$
$$+ \cos\psi \sin\psi k_y\left(k_{1z}^o + k_{oz}\right)\left(k_{1z}^o - k_{1z}^{eu}\right)\left(k_\rho^2 + k_{1z}^o k_{oz}\right) \tag{5.101}$$

References

1. J.K. Lee, and J.A. Kong, "Dyadic Green's functions for layered anisotropic medium," *Electromagnetics*, vol. 3, 111–130, 1983.
2. C.T. Tai, *Dyadic Green's Functions in Electromagnetic Theory*, IEEE Press, 1994.
3. C.M. Bender and S.A. Orszag, *Advanced Mathematical Methods for Scientists and Engineers*. New York: McGraw-Hill, 1978, pp. 280–302.
4. J.A. Kong, *Theory of Electromagnetic Waves*. New York: Wiley, 1975, pp. 59–62, 205–212.
5. R.W.P King and S.S. Sandler, "The electromagnetic field of a vertical electric dipole over the earth or sea," *IEEE Trans. Antennas Propag.*, vol. 42, no. 3, March 1994.
6. J.L. Tsalamengas and N.K. Uzunoglu, "Radiation from a dipole in the proximity of a gen eral anisotropic grounded layer," *IEEE Trans. Antennas Propag.*, vol. AP-33, no.2, pp. 165–172, 1985.
7. A. Eroglu, *Electromagnetic Wave Propagation and Radiation in Gyrotropic Medium*, Ph.D. dissertation, Dept. of Electrical Eng. and Computer Science, Syracuse University, 2004.

Chapter 6
Radiation in Gyrotropic Medium

The electromagnetic radiation from an electric Hertzian dipole in an electrically gyrotropic medium such as magnetically biased cold plasma, i.e. magnetoplasma, is very important for radio propagation problem in the ionosphere and for some problems in laboratory plasmas.

In this chapter, we study the far field radiation from an unbounded electrically gyrotropic medium when the applied external magnetic field is in z-direction. We use the method of dyadic Green's functions in spectral domain. Since the dyadic Green's function is expressed in terms of a single dyad, this facilitates the analytical calculation in comparison to the existing methods. The far field components are analytically derived using the method of steepest descents. Derivation of the analytical results is based on the assumption that the angle between the wave normal and the magnetic field $\overline{B}_0$ is known. It is shown that the acceptable range of these angles corresponding to individual wave types arises from the saddle point evaluation and can be determined from the transcendental equation numerically. The compact form of the dyadic Green's functions which involve a single dyad facilitates the analytical derivation of the far fields. The analytical derivation for the evaluation of the far fields for this type of medium is given in a systematic order. The investigation of the radiation characteristics using the CMA diagram gives an insight to a designer before the design for an antenna to determine the frequency of operation for which radiation characteristics can be determined. Hence, for a dispersive medium such as cold plasma or the ionosphere, the method outlined in this chapter can be used to determine frequency bands acceptable for the radiation fields.

6.1 Formulation of the Problem

The medium under consideration is an electrically gyrotropic medium such as cold plasma and is described by a relative permittivity tensor in dyadic form as

$$\bar{\bar{\varepsilon}} = \varepsilon_1(\bar{\bar{I}} - \hat{b}_0\hat{b}_0) + i\varepsilon_2(\hat{b}_0 \times \bar{\bar{I}}) + \varepsilon_3\hat{b}_0\hat{b}_0 \tag{6.1}$$

A. Eroglu, *Wave Propagation and Radiation in Gyrotropic and Anisotropic Media*,
DOI 10.1007/978-1-4419-6024-5_6,

or in the matrix form

$$\bar{\bar{\varepsilon}} = \begin{bmatrix} \varepsilon_1 & -i\varepsilon_2 & 0 \\ i\varepsilon_2 & \varepsilon_1 & 0 \\ 0 & 0 & \varepsilon_3 \end{bmatrix} \tag{6.2}$$

where

$$\varepsilon_1 = 1 - \frac{X}{1-Y^2} = 1 - \frac{{\omega_p}^2}{\omega^2 - {\omega_b}^2} \tag{6.3}$$

$$\varepsilon_2 = -\frac{XY}{1-Y^2} = -\frac{\omega_b {\omega_p}^2}{\omega(\omega^2 - {\omega_b}^2)} \tag{6.4}$$

$$\varepsilon_3 = 1 - X = 1 - \frac{{\omega_p}^2}{\omega^2} \tag{6.5}$$

and $\hat{b}_0 = \hat{z}$ is the direction of the magnetostatic field. The geometry of the medium is shown in Fig. 6.1 when the applied external magnetic field is in z-direction with an arbitrary direction of a wave vector $\bar{k}$.

We consider the problem of far field radiation from an unbounded gyroelectric medium such as cold plasma, when the dipole is located at the origin. The geometry of the problem is shown in Fig. 6.2.

The dyadic Green's function (DGF) of this structure when the impulse source is located at z = z' satisfies the following wave equation

$$\left[\nabla \times \nabla \times \bar{\bar{I}} - k_0^2 \bar{\bar{\varepsilon}}\right] \cdot \bar{\bar{G}}^e_{ee}(\bar{r}, \bar{r}') = i\omega\mu_0 \bar{\bar{I}} \delta(\bar{r} - \bar{r}') \tag{6.6}$$

with the radiation condition at $z = \pm\infty$. The superscript of the DGF, $\bar{\bar{G}}^e_{ee}(\bar{r}, \bar{r}')$, refers to the type of the gyrotropic medium, which is electric type in this case, and the first and the second subscripts show type of the DGF Green's function. The subscript "e" refers to the electric type and "m" refers to the magnetic type DGF.

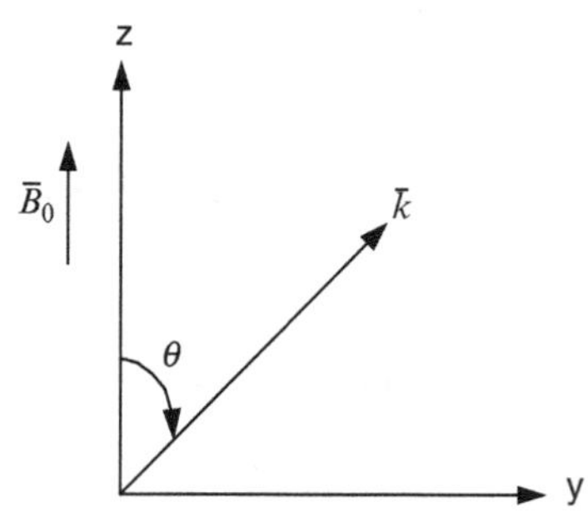

Fig. 6.1 Wave propagation in an electrically gyrotropic medium with an arbitrary direction of $\bar{k}$ and applied magnetostatic field $\bar{B}_0$

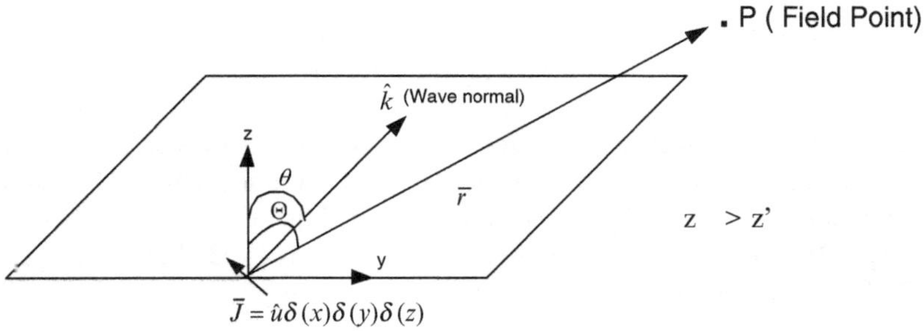

Fig. 6.2 The geometry of the problem when the dipole is located at the origin

The DGF, $\bar{\bar{G}}^e_{ee}(\bar{r},\bar{r}')$ given by (6.6) is derived in Chap. 4 when $z > z'$ as

$$\bar{\bar{G}}^e_{ee}(\bar{r},\bar{r}') = \frac{-\omega\mu_0}{8\pi^2} \times \int_{-\infty}^{\infty}\int_{-\infty}^{\infty} dk_x dk_y \left\{ \begin{array}{l} \dfrac{1}{k_0{}^2\varepsilon_3(k_{zI}{}^2 - k_{zII}{}^2)} \left[\dfrac{\alpha_I}{k_{zI}} \hat{e}_{nI}(k_{zI})\hat{e}^*_{nI}(k_{zI}) e^{i\bar{k}_I\cdot(\bar{r}-\bar{r}')} \right. \\ \left. - \dfrac{\alpha_{II}}{k_{zII}} \hat{e}_{nII}(k_{zII})\hat{e}^*_{nII}(k_{zII}) e^{i\bar{k}_{II}\cdot(\bar{r}-\bar{r}')} \right] \end{array} \right\} z > z' \tag{6.7}$$

The two dimensional form of the DGF in (6.7) is obtained after performing integral over k_z by assuming that the medium to be slightly lossy, i.e. $\mathrm{Im}k_z << \mathrm{Re}k_z, \mathrm{Im}k_z > 0$. This guarantees the radiation condition at $z = \pm\infty$. The parameters $\alpha_i, \hat{e}_{ni}(k_{zi}),\ i = I, II$ are derived in Chap. 4 and given by (4.114), (4.115) and $k_i,\ k_{zi},\ i = I, II$ are derived in Chap. 3 and given by (3.57), (3.58), (3.62) and (3.63), respectively.

6.2 Analytical Solution of Far Fields

In order to determine the far field radiation from a dipole in an unbounded gyro-electric medium, the far field approximated DGF of the problem in Fig. 6.2 has to be found for z > z'. The electric field $\bar{E}$ in the same region satisfies

$$\nabla \times \nabla \times \bar{E}(\bar{r}) - k_0^2\bar{\bar{\varepsilon}}.\bar{E}(\bar{r}) = i\omega\mu_0\bar{J}(\bar{r}) \tag{6.8}$$

The radiation field can be calculated by using the following relation between the electric field and the DGF:

$$\bar{E}(\bar{r}) = \int_V \bar{\bar{G}}^e_{ee}(\bar{r},\bar{r}').\ \bar{J}(\bar{r}')d^3\bar{r}' \tag{6.9}$$

where

$$\bar{J}(\bar{r}') = \hat{u}c\delta(x')\delta(y')\delta(z') \tag{6.10}$$

when the source is an electric Hertzian dipole. In (6.10), c is the current moment of the electric dipole and $\hat{u}$ is the orientation of the dipole. Under the far field approximation, the integral in (6.7) can be evaluated using the method of steepest descent [1, 2] as r $\to \infty$. We assume that the dipole is located at the origin, i.e. $\bar{r}' = 0$. We define the integral in (6.7) as

$$\bar{\bar{G}}^e_{ee}(\bar{r}, \bar{r}') = I_1 - I_2 \tag{6.11}$$

where

$$I_1 = \int\int_{-\infty}^{\infty} e^{i\bar{k}_I \cdot \bar{\rho}} \bar{\bar{F}}_I(\bar{k}_\rho) dk_x dk_y \tag{6.12}$$

$$I_2 = \int\int_{-\infty}^{\infty} e^{i\bar{k}_{II} \cdot \bar{\rho}} \bar{\bar{F}}_{II}(\bar{k}_\rho) dk_x dk_y \tag{6.13}$$

and

$$\bar{\bar{F}}_I(\bar{k}_\rho) = \frac{-\omega\mu_0}{8\pi^2} \frac{1}{k_{zI}} \frac{1}{k_0^2 \varepsilon_3 (k_{zI}^2 - k_{zII}^2)} \left[\alpha_I \hat{e}_{nI} \hat{e}^*_{nI} e^{ik_{zI} z}\right] \tag{6.14}$$

$$\bar{\bar{F}}_{II}(\bar{k}_\rho) = \frac{-\omega\mu_0}{8\pi^2} \frac{1}{k_{zII}} \frac{1}{k_0^2 \varepsilon_3 (k_{zI}^2 - k_{zII}^2)} \left[\alpha_{II} \hat{e}_{nII} \hat{e}^*_{nII} e^{ik_{zII} z}\right] \tag{6.15}$$

Please note that for simplicity in the notation we assumed $\bar{\rho}$ for $\bar{\rho} - \bar{\rho}'$ and z for $z - z'$. Since the calculations for I_1 and I_2 are similar, we will show only the analytical calculations for I_1. We first represent the integral in (6.12) in cylindrical wave form by the following transformations.

$$\bar{k}_\rho = \hat{x} k_\rho \cos\alpha + \hat{y} k_\rho \sin\alpha \tag{6.16}$$

$$\bar{\rho} = \hat{x}\rho\cos\phi + \hat{y}\rho\sin\phi \tag{6.17}$$

$$dk_x dk_y = k_\rho dk_\rho d\alpha \tag{6.18}$$

This transformation can be graphically shown in Fig. 6.3. It is also clear from Fig. 6.3 that

$$\bar{k}_\rho . \bar{\rho} = k_\rho \ \rho \ \cos(\alpha - \phi) \tag{6.19}$$

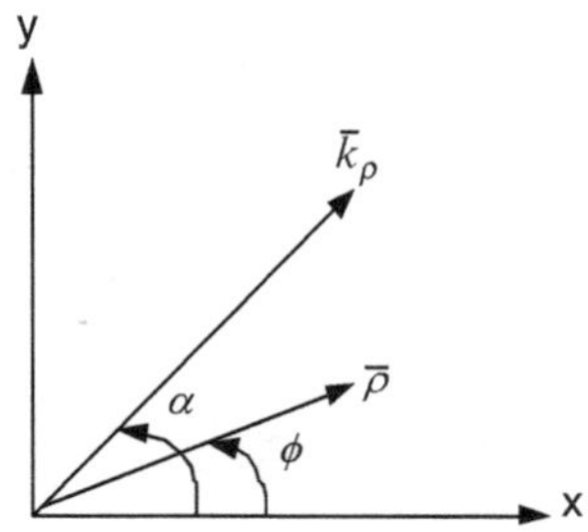

Fig. 6.3 k_ρ and ρ vectors on the xy plane

Substituting (6.16)–(6.18) into (6.12) gives

$$I_1 = \int_0^\infty k_\rho \bar{\bar{F}}_I(k_\rho) dk_\rho \int_0^{2\pi} d\alpha e^{ik_\rho \rho \cos(\alpha - \phi)} \tag{6.20}$$

Using the following integral identity for Bessel functions [3]:

$$J_0(k_\rho \rho) = \frac{1}{2\pi} \int_0^{2\pi} d\alpha e^{ik_\rho \rho \cos(\alpha - \phi)} \tag{6.21}$$

The integral in (6.20) can be written as

$$I_1 = 2\pi \int_0^\infty k_\rho \bar{\bar{F}}_I(k_\rho) J_0(k_\rho \rho) dk_\rho \tag{6.22}$$

Using the relation between the Bessel functions [4], $J_0(k_\rho \rho)$ can be written as

$$J_0(k_\rho \rho) = \frac{1}{2}\left[H_0^{(1)}(k_\rho \rho) + H_0^{(2)}(k_\rho \rho)\right] \tag{6.23}$$

Substituting (6.23) into (6.22) gives

$$I_1 = \pi \int_0^\infty k_\rho \bar{\bar{F}}_I(k_\rho) \left[H_0^{(1)}(k_\rho \rho) + H_0^{(2)}(k_\rho \rho)\right] dk_\rho$$

or

$$I_1 = \pi \int_0^\infty k_\rho \bar{\bar{F}}_I(k_\rho) \left[H_0^{(1)}(k_\rho \rho)\right] dk_\rho + \pi \int_0^\infty k_\rho \bar{\bar{F}}_I(k_\rho) \left[H_0^{(2)}(k_\rho \rho)\right] dk_\rho \tag{6.24}$$

Consider the second part of the integral in (6.24) and change the variable of the integration as

$$k_\rho = e^{-i\pi} k'_\rho \tag{6.25}$$

Then,

$$dk_\rho = e^{-i\pi} dk_\rho' = -dk_\rho' \text{ and as } k_\rho \to \infty, k_\rho' \to -\infty \tag{6.26}$$

Hence the second part of the integral can be written as

$$\pi \int_0^{-\infty} k_{\rho'} \bar{\bar{F}}_I(k_{\rho'}) \left[H_0^{(2)}(e^{-i\pi} k_{\rho'} \rho) \right] dk_{\rho'} \tag{6.27}$$

Using the reflection formula to relate the first and the second kind Hankel functions [5]

$$H_0^{(2)}(e^{-i\pi}\lambda) = -H_0^{(1)}(\lambda) \tag{6.28}$$

and substituting it back into (6.27) gives

$$-\pi \int_0^{-\infty} k_{\rho'} \bar{\bar{F}}_I(k_{\rho'}) \left[H_0^{(1)}(k_{\rho'} \rho) \right] dk_{\rho'}$$

or

$$\pi \int_{-\infty}^{0} k_{\rho'} \bar{\bar{F}}_I(k_{\rho'}) \left[H_0^{(1)}(k_{\rho'} \rho) \right] dk_{\rho'} \tag{6.29}$$

When we combine (6.29) and (6.24), we can express (6.20) in terms of the Hankel function of the first kind which is chosen to represent an outgoing wave using the "*i*" notation, as

$$I_1 = \pi \int_{-\infty}^{\infty} k_\rho \bar{\bar{F}}_I(k_\rho) H_0^{(1)}(k_\rho \rho) dk_\rho \tag{6.30}$$

Substituting (6.14) into (6.30) gives

$$I_1 = \int_{-\infty}^{\infty} g(k_\rho) e^{ik_{zI}z} \frac{k_\rho}{k_{zI}} H_0^{(1)}(k_\rho \rho) dk_\rho \tag{6.31}$$

where

$$g(k_\rho) = \frac{-\omega\mu_0}{8\pi k_0^2 \varepsilon_3} \left(\frac{\alpha_I \hat{e}_{nI} \hat{e}_{nI}^*}{k_{zI}^2 - k_{zII}^2} \right) \tag{6.32}$$

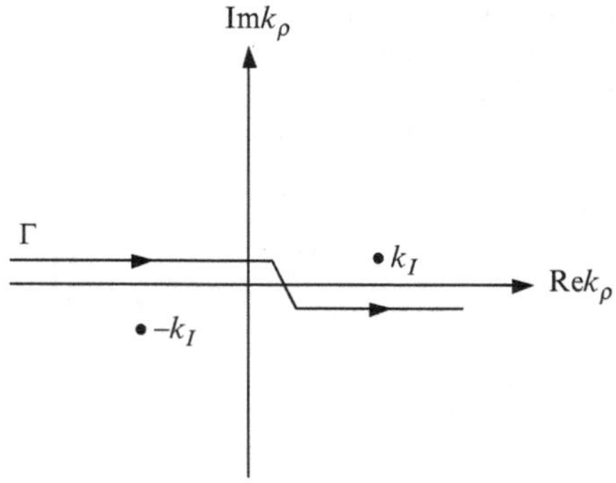

Fig. 6.4 The Sommerfeld integration path on complex k_ρ plane

In (6.31) inside the integral, a plane wave propagating in the z- direction is in multiplication with the Henkel function of first kind which represents a cylindrical wave propagation in ρ direction. They are in multiplication with an amplitude factor $g(k_\rho)$. The integral in (6.31) has a branch point singularity at $k_\rho = \pm k_I$ since $k_{zI} = \left[k_I^2 - k_\rho^2\right]^{1/2}$ is a double valued function of k_ρ. Also $H_0^{(1)}(\lambda)$ has a branch point singularity when $\lambda = 0$. Hence the integral in (6.31) is undefined unless we identify the path of integration. The typical path of integration for the integral type in (6.31) is shown in Fig. 6.4.

As seen from Fig. 6.4, the branch point singularities $\pm k_I$ have been displaced slightly from the real axis to signify the presence of loss to satisfy the radiation condition. The path of integration Γ is also known as Sommerfeld path of integration. To make a definition of a double valued function k_{zI} unique, a two-sheeted complex k_ρ is necessary, with branch cuts providing the means of passing from one Riemann sheet to the other. The selection of the branch cuts is arbitrary but determines the disposition of those regions of the complex k_ρ plane in which $\mathrm{Im}k_{zI} > 0$ or $\mathrm{Im}k_{zI} < 0$. To identify the branch cuts that we need in the evaluation of the integral in (6.31) we can consider the following. Let us define

$$\left[k_I - k_\rho\right] = \left|k_I - k_\rho\right| e^{i\alpha} \text{ and } \left[k_I + k_\rho\right] = \left|k_I + k_\rho\right| e^{i\beta}, \alpha, \beta \text{ are real} \tag{6.33}$$

The angles α and β are selected such that $\alpha = 0$ and $\beta = 0$ when k_ρ is real and $\left|k_\rho\right| < k_I$ on the top Riemann sheet. From (6.33)

$$\sqrt{k_I^2 - k_\rho^2} = \left|\sqrt{k_I^2 - k_\rho^2}\right| e^{i\frac{(\alpha+\beta)}{2}} \tag{6.34}$$

When k_ρ is real and $\left|k_\rho\right| < k_I$ where k_I is assumed to be real, the wave along z-direction is propagating and hence the propagation constant k_{zI} is real and positive. Thus we require a definition of k_{zI} such that

$$\sqrt{k_I^2 - k_\rho^2} > 0, -k_I < k_\rho < k_I \tag{6.35}$$

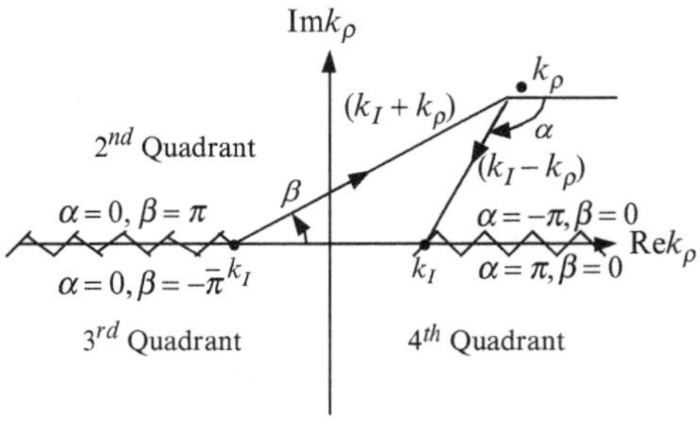

Fig. 6.5 $\mathrm{Im}k_{zI} > 0$ in region 2 and region 4, $\mathrm{Im}k_{zI} < 0$ in region 1 and region 3, $\mathrm{Re}k_{zI} > 0$ on the entire top sheet

With the requirements defined above and (6.33) and (6.34) and to satisfy (6.35), it is required that

$$(\alpha + \beta) = 0 \quad \text{When} \quad -k_I < k_\rho < k_I \tag{6.36}$$

The selection of the branch cut is illustrated in Fig. 6.5 using the requirements defined above.

With the angles of α and β defined as shown, it is clear that the requirement in Equation (6.36) is met since $\alpha = \beta = 0$ when $-k_I < k_\rho < k_I$. Hence this position of the branch cuts identifies the ranges of α and β on the top sheet to

$$-\pi < \alpha < \pi \quad \text{and} \quad -\pi < \beta < \pi \tag{6.37}$$

With the usual convention, we assumed that an angle is negative if it is clockwise and positive if it is measured counterclockwise. As shown in Fig. 6.5, in the 1st and 3^{rd}quadrants of k_ρ plane,

$$-\pi < \alpha + \beta < 0 \tag{6.38}$$

hence

$$\mathrm{Im}k_{zI} = \left|\sqrt{k_I^2 - k_\rho^2}\right| \sin(\frac{\alpha + \beta}{2}) < 0 \tag{6.39}$$

$$\mathrm{Re}k_{zI} = \left|\sqrt{k_I^2 - k_\rho^2}\right| \cos(\frac{\alpha + \beta}{2}) > 0 \tag{6.40}$$

While in the 2nd and 4^{th}quadrants of k_ρ plane,

$$0 < \alpha + \beta < \pi \tag{6.41}$$

hence

$$\mathrm{Im}k_{zI} = \left|\sqrt{k_I^2 - k_\rho^2}\right| \sin(\frac{\alpha + \beta}{2}) > 0 \tag{6.42}$$

$$\text{Re}k_{zI} = \left|\sqrt{k_I^2 - k_\rho^2}\right| \cos(\frac{\alpha+\beta}{2}) > 0 \tag{6.43}$$

In the integral given by (6.31), the values of k_{zI} have to be such that $\text{Im}(k_{zI}) > 0$ over the path of integration in order to satisfy the radiation condition. In the above discussion, we have assumed k_I to be real. However, we can slightly displace $\pm k_I$ from the real axis to signify the presence of loss to satisfy the radiation condition. The branch cut is defined when $\text{Im}(k_{zI}) = 0$. This requirement also sets a boundary between the quadrants since $\text{Im}(k_{zI}) > 0$ in the 2nd and 4^{th}quadrants of the k_ρ plane and $\text{Im}(k_{zI}) < 0$ in the 1st and 3^{rd}quadrants of the k_ρ plane. To find such a branch cut we assume,

$$k_I = k_{rI} + ik_{iI} \quad \text{and} \quad k_\rho = k_{r\rho} + ik_{i\rho} \tag{6.44}$$

Then

$$k_{zI} = \left[k_I^2 - k_\rho^2\right]^{1/2} = \left[k_{rI}^2 - k_{iI}^2 + i2k_{rI}k_{iI} - k_{r\rho}^2 + k_{i\rho}^2 - i2k_{r\rho}k_{i\rho}\right]^{1/2} \tag{6.45}$$

For $\text{Im}(k_{zI}) = 0$ or in other words for k_{zI} to be real,

$$k_{r\rho}k_{i\rho} = k_{rI}k_{iI}, \quad \text{and} \quad k_{r\rho}^2 - k_{i\rho}^2 < k_{rI}^2 - k_{iI}^2 \tag{6.46}$$

Equation (6.46) defines hyperbolas that pass through the branch points $\pm k_I$ as shown in Fig. 6.6. Hence we can construct our complete integration path in k_ρ plane as shown in Fig. 6.7 using the results that we obtained.

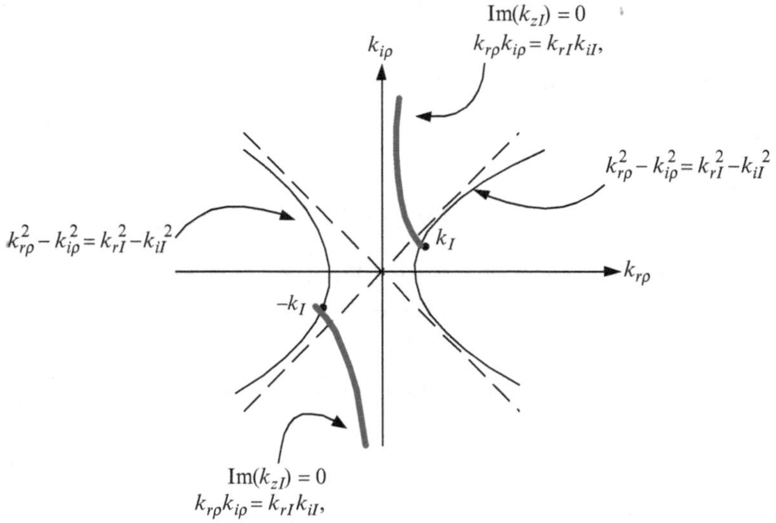

Fig. 6.6 Construction of the branch cuts

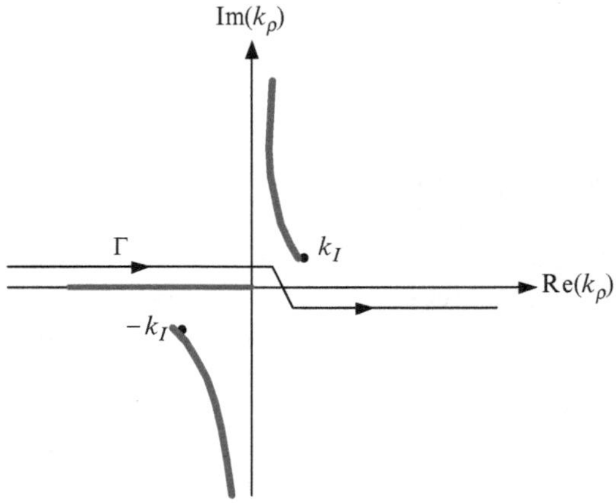

Fig. 6.7 The complete integration path for the integral in (6.29) in the complex k_ρ plane

To facilitate the solution for the integral in (6.31), at this point, it is convenient to transform from the complex wavenumber variable k_ρ to complex angle variable β. The transformation can be done as follows.

$$k_\rho = k_I \sin\beta \tag{6.47}$$

Then, we also have

$$k_{zI} = k_I \cos\beta \tag{6.48}$$

where β is the complex angle between the wave normal and the z-axis and is expressed as

$$\beta = \beta_r + i\beta_i \tag{6.49}$$

Also we introduce the following transformations to polar coordinates as follows.

$$\rho = r\sin\Theta \tag{6.50}$$

$$z = r\cos\Theta \tag{6.51}$$

This transformation can be illustrated in Fig. 6.8.

Substituting (6.47)–(6.51) into (6.31), we obtain

$$I_1 = \int_P g(k_I\sin\beta)\tan\beta H_0^{(1)}(k_I r\sin\beta\sin\Theta)(k_I'\sin\beta + k_I\cos\beta)e^{ik_I r\cos\beta\cos\Theta}d\beta \tag{6.52}$$

where $k_I' = \frac{\partial k_I}{\partial\beta}$. As seen in (6.52), the branch point pair at $\pm k_I$ is removed by the transformation that we suggested in (6.47)–(6.51). The new integral path P due to

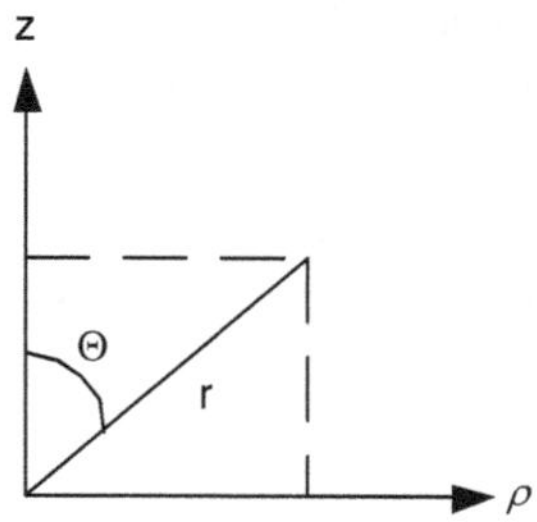

Fig. 6.8 Polar transformation

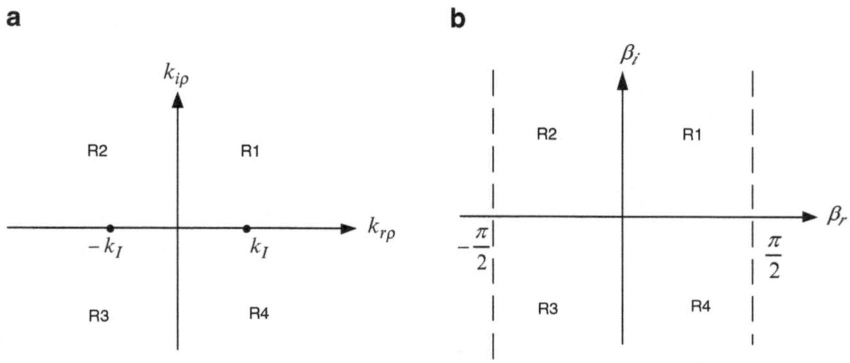

Fig. 6.9 (**a**) Regions in k_ρ plane. (**b**) Top Riemann sheet on k_ρ plane is mapped to β plane

this transformation or mapping can be explained as follows. Substitute (6.49) into (6.47) and separate (6.47) into its real and imaginary parts and assume that k_I is real. Then

$$k_{r\rho} = k_I \sin\beta_r \cosh\beta_i \quad \text{and} \quad k_{i\rho} = k_I \cos\beta_r \sinh\beta_i \tag{6.53}$$

where

$$k_\rho = k_{r\rho} + ik_{i\rho} \quad \text{and} \quad \beta = \beta_r + i\beta_i \tag{6.54}$$

The four quadrants in the complex k_ρ plane map into corresponding regions in the complex β plane as identified by (6.53) and (6.54) as shown in Fig. 6.9. Hence the top sheet of the Reimann surface in the k_ρ plane maps into the region $-\frac{\pi}{2} < \beta_r < \frac{\pi}{2}$ in the β plane [6]. To identify the path P explicitly, we investigate the convergence of the integral in (6.31) in the neighborhood of infinity by analyzing the behavior of the term in the exponent ik_{zI} in the β plane as follows.

$$ik_{zI} = ik_I \cos\beta = k(i\cos\beta_r \cosh\beta_i + \sin\beta_r \sinh\beta_i) \tag{6.55}$$

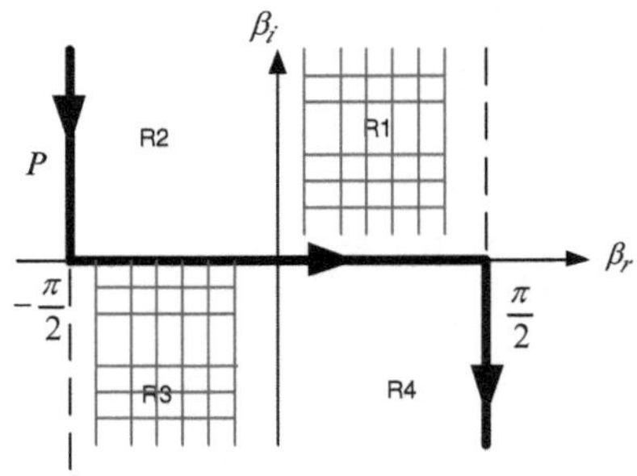

Fig. 6.10 Construction of the integral path P in the β plane

As
$\beta_i \to \infty$, Real Part is positive, if $\sin\beta_r > 0 \to 0 < \beta_r < \frac{\pi}{2}$
$\beta_i \to \infty$, Real Part is negative, if $\sin\beta_r < 0 \to -\frac{\pi}{2} < \beta_r < 0$
$\beta_i \to -\infty$, Real Part is positive, if $\sin\beta_r < 0 \to -\frac{\pi}{2} < \beta_r < 0$
$\beta_i \to -\infty$, Real Part is negative, if $\sin\beta_r > 0 \to 0 < \beta_r < \frac{\pi}{2}$
When the real part is positive, the integral diverges exponentially. They are shown with the shaded areas in Fig. 6.10. These regions are the 1st and 3rd regions in the β plane. For the integral in (6.31) to be convergent, the real part has to go to $-\infty$. From (6.55)

$$\sin\beta_r \sinh\beta_i \to -\infty \text{ when } \beta_r = -\frac{\pi}{2} \text{ and } \beta_i \to \infty \;\; \beta_r = \frac{\pi}{2} \text{ and } \beta_i \to -\infty \qquad (6.57)$$

Hence, the integral given by the (6.31) is convergent under the conditions outlined above. The integration path P in Fig. 6.10 is constructed to meet the requirements in (6.56) and (6.57).

When we combine the integral path that we constructed in this plane with the branch cut requirements shown in Fig. 6.7, we obtain the complete integration path P in the β plane as shown in Fig. 6.11.

From the condition that $k_\rho \rho >> 1$ it necessarily follows that $k_I r >> 1$ in the integral in (6.52). Then we can employ large argument approximation of $H_0^{(1)}$ and take its first asymptotic approximation as

$$H_0^{(1)}(k_I r \sin\beta \sin\Theta) = \sqrt{\frac{2}{\pi\, k_I r \sin\beta \sin\Theta}}\, e^{i(k_I r \sin\beta \sin\Theta - \frac{\pi}{4})} \qquad (6.58)$$

When (6.58) is inserted into the integral in (6.52), we obtain

$$I_1 = \left(\frac{-\omega\mu_0}{8\pi\, k_0^2 \varepsilon_3}\right) e^{-i\frac{\pi}{4}} \sqrt{\frac{2}{\pi\, r \sin\Theta}} \int_P h(\beta) e^{ik_I r \cos(\beta - \Theta)} d\beta \qquad (6.59)$$

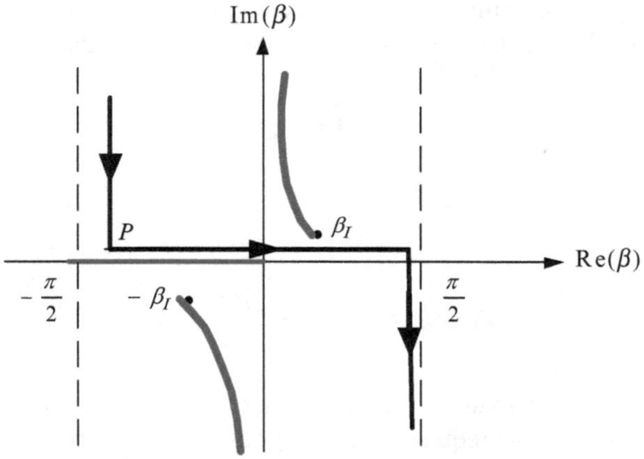

Fig. 6.11 The complete integration path P for the integral in (6.46) in the β plane

where

$$h(\beta) = \frac{(k_I' \sin\beta + k_I \cos\beta)}{\cos\beta} \sqrt{\frac{\sin\beta}{k_I}} \left(\frac{\alpha_I \hat{e}_{nI} \hat{e}_{nI}^*}{k_{zI}^2 - k_{zII}^2} \right) \tag{6.60}$$

The integral in (6.59) can be solved by using the method of the steepest descent. The integration path P can be deformed into the steepest descent path by setting the imaginary part of the exponent to be equal to its value at the saddle point. The integral given by (6.59) can be written in the following form.

$$I_{s1}(\Omega) = \int_P h(\beta) e^{\Omega q(\beta)} d\beta \tag{6.61}$$

where

$$\Omega \to r \tag{6.62}$$

$$q(\beta) = ik_I \ \cos(\beta - \Theta) \tag{6.63}$$

The integration path P ends at infinity and the large parameter Ω is positive. The functions $h(\beta)$ and $q(\beta)$ are analytic functions of β along the path of integration P. We can transform the integral in (6.61) into a canonical form wherein the function $q(\beta)$ is replaced by another function, a polynomial that has the simplest form of the relevant saddle point arrangement at β_s [6]. The transformation will be phrased in terms of a new variable s and the polynomial $\tau(s)$:

$$\tau(s) = q(\beta) \tag{6.64}$$

The point β_s in the complex β plane is chosen to correspond to $s = 0$ in the complex $s$plane. Hence, (6.61) leads to the transformed integral

$$I_{s1}(\Omega) = \int_{P'} G(s)e^{\Omega\tau(s)}ds \tag{6.65}$$

where

$$G(s) = h(\beta(s))\frac{d\beta}{ds} \text{ and } \frac{d\beta}{ds} = \frac{\tau'(s)}{q'(\beta)} \tag{6.66}$$

In our case we are solving the problem, assuming that the saddle point is an isolated saddle point. This requires that $q'(\beta)$ has a simple zero at β_s and no other zero near β_s. When the saddle point is an isolated saddle point,

$$q(\beta) = \tau(s) = q(\beta_s) - s^2 \tag{6.67}$$

or

$$-s^2 = q(\beta) - q(\beta_s) \tag{6.68}$$

The steepest descent path P' in the s-plane, along which $\text{Im}\tau(s) = \text{constant}$, is clearly the real s-axis. Thus the integral in (6.62) becomes

$$I_{s1}(\Omega) = e^{\Omega q(\beta_s)} \int_{-\infty}^{\infty} G(s)e^{-\Omega s^2}ds \tag{6.69}$$

where $G(s)$ is given by (6.63). We assume that $G(s)$ is a smooth function of s; then the integral I_{s1} is dominated by the exponential $e^{-\Omega s^2}$, which possesses the maximum value at $s = 0$ and decreases rapidly on both sides of s. Hence the saddle point in the β plane corresponds to $s = 0$ in the s plane. We expand $G(s)$ into a powers series near the saddle point $s = 0$ as

$$G(s) = \sum_{m=0}^{\infty} A_m s^m \tag{6.70}$$

and obtain the asymptotic series for $I_{s1}(\Omega)$, we use the formulas [2]

$$\int_{-\infty}^{\infty} s^{2m+1}e^{-\Omega s^2}ds = 0 \tag{6.71}$$

$$\int_{-\infty}^{\infty} s^{2m}e^{-\Omega s^2}ds = (-1)^m\frac{d^m}{d\Omega^m}\sqrt{\frac{\pi}{\Omega}} = \frac{(2m)!}{m!2^{2m}\Omega^m}\sqrt{\frac{\pi}{\Omega}} \tag{6.72}$$

When we substitute (6.70) and (6.72) into the (6.69), we obtain

$$I_{s1}(\Omega) = e^{\Omega q(\beta_s)} \sum_{m=0}^{\infty} A_m \frac{(2m)!}{m! 2^{2m} \Omega^m} \sqrt{\frac{\pi}{\Omega}} \tag{6.73}$$

We need to determine the expansion coefficient A_m. For this, we first expand $\beta(s)$ and (6.68) around $s = 0$ where the integral to be evaluated:

$$\beta(s) - \beta_s = \sum_{n=1}^{\infty} a_n s^n \tag{6.74}$$

and

$$-s^2 = \frac{1}{2!} q''(\beta_s)(\beta - \beta_s)^2 + \frac{1}{3!} q'''(\beta_s)(\beta - \beta_s)^3 + \frac{1}{4!} q''''(\beta_s)(\beta - \beta_s)^4 + \ldots \tag{6.75}$$

When we substitute (6.74) into (6.75), we obtain

$$\begin{aligned} -s^2 &= \frac{1}{2} q''(\beta_s)[a_1^2 s^2 + 2a_1 a_2 s^3 + (a_2^2 + 2a_1 a_3)s^4 + \ldots] \\ &+ \frac{1}{6} q'''(\beta_s)[a_1^3 s^3 + 3a_1^2 a_2 s^4 + \ldots] + \end{aligned} \tag{6.76}$$

To determine a_n, we compare the coefficients of s^2, s^3, s^4, etc. on the left hand side and the right hand side of (6.76). Hence, we find the coefficients as

$$a_1 = \sqrt{\frac{-2}{q''(\beta_s)}} \tag{6.77}$$

$$a_2 = -\frac{q'''(\beta_s)}{6q''(\beta_s)} a_1^2 \tag{6.78}$$

From (6.66) and (6.74), $G(s)$ can be written as

$$G(s) = \frac{d\beta}{ds} \sum_{k=0}^{\infty} \frac{1}{k!} h^{(k)}(\beta_s)(\beta - \beta_s)^k \tag{6.79}$$

or

$$G(s) = \left[\sum_{m=1}^{\infty} m a_m s^{m-1}\right] \sum_{k=0}^{\infty} \frac{1}{k!} h^{(k)}(\beta_s) \left[\sum_{n=1}^{\infty} a_n s^n\right]^k \tag{6.80}$$

Using the coefficients a_n given by (6.77) and (6.78), we can express $G(s)$ as

$$G(s) = h(\beta_s)(a_1 + 2a_2 s + 3a_3 s^2 + ...) + h'(\beta_s)(a_1 s + 2a_2 s^2 + ...) \\ (a_1 + 2a_2 s + ...) + ... \tag{6.81}$$

Comparison of the coefficients in (6.81) and (6.70), we find the coefficients A_m as

$$A_0 = a_1 h(\beta_s) = \sqrt{\frac{-2}{q''(\beta_s)}} h(\beta_s) \tag{6.82}$$

$$A_2 = 3a_3 h(\beta_s) + 3a_1 a_2 h'(\beta_s) + \frac{1}{2} a_1^3 h''(\beta_s) \tag{6.83}$$

The odd numbered coefficients A_m are zero from (6.71). Substituting (6.82) and (6.83) into (6.73) and using (6.77) and (6.78), we obtain the integral $I_{s1}(\Omega)$ as

$$I_{s1}(\Omega) = h(\beta_s) e^{\Omega q(\beta_s)} \sqrt{\frac{-2\pi}{\Omega q''(\beta_s)}} \left\{ 1 + \frac{1}{2\Omega q''(\beta_s)} \left[\frac{q'''(\beta_s)}{q''(\beta_s)} \frac{h'(\beta_s)}{h(\beta_s)} + \frac{1}{4} \frac{q''''(\beta_s)}{q''(\beta_s)} + ... \right.\right. \\ \left.\left. - \frac{5}{2} \left(\frac{q'''(\beta_s)}{q''(\beta_s)} \right)^2 - \frac{h''(\beta_s)}{h(\beta_s)} + ... \right] \right\} \tag{6.84}$$

Since the dominant term is the first term in the integral given by (6.84), it can be approximated by

$$I_{s1}(\Omega) \approx h(\beta_s) \sqrt{\frac{-2\pi}{q''(\beta_s)\Omega}} e^{\Omega q(\beta_s)}, \Omega \to \infty \tag{6.85}$$

We need to identify the saddle point β_s and $q''(\beta_s)$ in (6.85). At the saddle point $\beta = \beta_s$,

$$q'(\beta) = 0 \tag{6.86}$$

$$q'(\beta)_{\beta=\beta_s} = k_I'(\beta_s) \cos(\beta_s - \Theta) - k_I(\beta_s) \sin(\beta_s - \Theta) = 0 \tag{6.87}$$

Then,

$$\frac{k_I'(\beta_s)}{k_I(\beta_s)} = \tan(\beta_s - \Theta) \tag{6.88}$$

Unfortunately, there is no analytical solution for β_s in (6.88). However, the values of β_s satisfying (6.88) can be calculated numerically for the given observation angle Θ. By using (6.63), we can find $q''(\beta_s)$ as

$$q''(\beta_s) = i\left[\cos(\beta_s - \Theta)[k_I''(\beta_s) - k_I(\beta_s)] - 2k_I'(\beta_s)\sin(\beta_s - \Theta)\right] \tag{6.89}$$

When (6.89) is substituted into (6.85) by replacing $\Omega \rightarrow r$ we obtain

$$I_{s1}(r) \approx h(\beta_s) \times \sqrt{\frac{2\pi}{[\cos(\beta_s - \Theta)[k_I''(\beta_s) - k_I(\beta_s)] - 2k_I'(\beta_s)\sin(\beta_s - \Theta)]\, r}}\, e^{ik_I r\cos(\beta_s - \Theta)} e^{i\frac{\pi}{4}}, r \rightarrow \infty \tag{6.90}$$

Substituting (6.90) back into (6.59) gives,

$$I_1 = \left(\frac{-\omega\mu_0}{8\pi\, k_0^2 \varepsilon_3}\right) e^{-i\frac{\pi}{4}} \sqrt{\frac{2}{\pi\, r \sin\Theta}}\, h(\beta_s) \times \sqrt{\frac{2\pi}{[\cos(\beta_s - \Theta)[k_I''(\beta_s) - k_I(\beta_s)] - 2k_I'(\beta_s)\sin(\beta_s - \Theta)]\, r}}\, e^{ik_I r\cos(\beta_s - \Theta)} e^{i\frac{\pi}{4}}$$

or

$$I_1 = \left(\frac{-\omega\mu_0}{4\pi\, k_0^2 \varepsilon_3}\right) \sqrt{\frac{1}{\sin\Theta}}\, h(\beta_s) \times \sqrt{\frac{1}{[\cos(\beta_s - \Theta)[k_I''(\beta_s) - k_I(\beta_s)] - 2k_I'(\beta_s)\sin(\beta_s - \Theta)]}}\, \frac{e^{ik_I r\cos(\beta_s - \Theta)}}{r} \tag{6.91}$$

where $h(\beta_s)$ is defined by (6.60) as

$$h(\beta_s) = \frac{(k_I'(\beta_s)\sin\beta_s + k_I(\beta_s)\cos\beta_s)}{\cos\beta_s} \sqrt{\frac{\sin\beta_s}{k_I(\beta_s)}} \left(\frac{\alpha_I \hat{e}_{nI}\hat{e}_{nI}^*}{k_{zI}^2 - k_{zII}^2}\right)_{\beta=\beta_s} \tag{6.92}$$

We can simplify (6.92) using (6.88) as

$$h(\beta_s) = \frac{\cos\Theta}{\cos(\beta_s)\cos(\beta_s - \Theta)} \sqrt{\sin\beta_s k_I(\beta_s)} \left(\frac{\alpha_I \hat{e}_{nI}\hat{e}_{nI}^*}{k_{zI}^2 - k_{zII}^2}\right)_{\beta=\beta_s} \tag{6.93}$$

From (3.62) and (3.63),

$$(k_{zI}^2 - k_{zII}^2) = \left[\frac{k_\rho^4(\beta_s)}{k_0^4}(\varepsilon_1 - \varepsilon_3)^2 + 4{\varepsilon_2}^2\varepsilon_3[\varepsilon_3 - \frac{k_\rho^4(\beta_s)}{k_0^2}]\right]^{1/2}\frac{{k_0}^2}{\varepsilon_{33}}$$

or

$$(k_{zI}^2 - k_{zII}^2) = \left[\frac{k_I^4(\beta_s)}{k_0^4}\sin^4(\beta_s)(\varepsilon_1 - \varepsilon_3)^2 + 4{\varepsilon_2}^2\varepsilon_3[\varepsilon_3 - \frac{k_I^2(\beta_s)}{k_0^2}\sin^2(\beta_s)]\right]^{1/2}\frac{{k_0}^2}{\varepsilon_3} \tag{6.94}$$

Substituting (6.93), (6.94) into (6.91) gives

$$\begin{aligned} I_1 = &\left(\frac{-\omega\mu_0}{4\pi k_0^4}\right)\frac{e^{ik_I r\cos(\beta_s-\Theta)}}{r} \\ &\times\left\{\left[\frac{\cos\Theta}{\sqrt{\sin\Theta}}\frac{1}{\cos(\beta_s)\cos(\beta_s-\Theta)}\left(\alpha_I\hat{e}_{nI}\hat{e}_{nI}^*\right)_{\beta=\beta_s}\right]\right. \\ &\times\left[\frac{\sin\beta_s k_I(\beta_s)}{[\cos(\beta_s - \Theta)[k_I''(\beta_s) - k_I(\beta_s)] - 2k_I'(\beta_s)\sin(\beta_s - \Theta)]}\right]^{1/2} \\ &\times\left.\left[\frac{k_I^4(\beta_s)}{k_0^4}\sin^4(\beta_s)(\varepsilon_1 - \varepsilon_3)^2 + 4{\varepsilon_2}^2\varepsilon_3[\varepsilon_3 - \frac{k_I^2(\beta_s)}{k_0^2}\sin^2(\beta_s)]\right]^{-1/2}\right\} \end{aligned} \tag{6.95}$$

Repeating the same procedure for the integral I_2 in (6.13), we obtain

$$\begin{aligned} I_1 = &\left(\frac{-\omega\mu_0}{4\pi k_0^4}\right)\frac{e^{ik_{II}\, r\cos(\beta_s-\Theta)}}{r} \\ &\times\left\{\left[\frac{\cos\Theta}{\sqrt{\sin\Theta}}\frac{1}{\cos(\beta_s)\cos(\beta_s-\Theta)}\left(\alpha_{II}\hat{e}_{nII}\hat{e}_{nII}^*\right)_{\beta=\beta_s}\right]\right. \\ &\times\left[\frac{\sin\beta_s k_{II}(\beta_s)}{[\cos(\beta_s - \Theta)[k_{II}''(\beta_s) - k_{II}(\beta_s)] - 2k_{II}'(\beta_s)\sin(\beta_s - \Theta)]}\right]^{1/2} \\ &\times\left.\left[\frac{k_{II}^4(\beta_s)}{k_0^4}\sin^4(\beta_s)(\varepsilon_1 - \varepsilon_3)^2 + 4{\varepsilon_2}^2\varepsilon_3[\varepsilon_3 - \frac{k_{II}^2(\beta_s)}{k_0^2}\sin^2(\beta_s)]\right]^{-1/2}\right\} \end{aligned} \tag{6.96}$$

Hence the far field approximated DGF $\bar{\bar{G}}_{ee}^e(\bar{r},\bar{r}')$for an electrically gyrotropic medium is obtained by substituting (6.95) and (6.96) into (6.11).

The radiation fields can be found by substituting (6.11) into (6.9) as follows.

$$\bar{E}(\bar{r}) = \int_v \bar{\bar{G}}_{ee}^e(\bar{r},\bar{r}').\bar{J}(\bar{r}')d\bar{r}' = \bar{\bar{G}}_{ee}^e(\bar{r},\bar{r}')\Big|_{x'=0,\ y'=0,\ z'=0}\bullet\hat{u}c \tag{6.97}$$

Hence, (6.97) can be written as

$$\overline{E}(\overline{r}) = \left(\frac{-\omega\mu_0}{4\pi\ k_0^4}\right)\frac{e^{ik_{I,II}r\cos(\beta_s-\Theta)}}{r}$$
$$\times\left\{\left[\frac{\cos\Theta}{\sqrt{\sin\Theta}}\frac{1}{\cos(\beta_s)\cos(\beta_s-\Theta)}\left(\alpha_{I,II}\hat{e}_{nI,II}\hat{e}^*_{nI,II}\right)_{\beta=\beta_s}\right]\cdot\right.$$
$$\times\left[\frac{\sin\beta_s k_{I,II}(\beta_s)}{\left[\cos(\beta_s - \Theta)[k''_{I,II}(\beta_s) - k_{I,II}(\beta_s)] - 2k'_{I,II}(\beta_s)\sin(\beta_s-\Theta)\right]}\right]^{1/2}$$
$$\left.\times\left[\frac{k^4_{I,II}(\beta_s)}{k_0^4}\sin^4(\beta_s)(\varepsilon_1-\varepsilon_3)^2+4{\varepsilon_2}^2\varepsilon_3[\varepsilon_3-\frac{k^2_{I,II}(\beta_s)}{k_0^2}\sin^2(\beta_s)]\right]^{-1/2}\right\}$$
$$\bullet\ \hat{u}c|_{x'=0,y'=0,z'=0} \tag{6.98}$$

We define the following variables to simplify the expressions for the radiation fields.

$$E = \left[\frac{\cos\Theta}{\sqrt{\sin\Theta}}\frac{1}{\cos(\beta_s)\cos(\beta_s-\Theta)}\right] \tag{6.99}$$

$$R = \left[\frac{\sin\beta_s k_{I,II}(\beta_s)}{\left[\cos(\beta_s - \Theta)[k''_{I,II}(\beta_s) - k_{I,II}(\beta_s)] - 2k'_{I,II}(\beta_s)\sin(\beta_s-\Theta)\right]}\right]^{1/2} \tag{6.100}$$

$$O = \left[\frac{k^4_{I,II}(\beta_s)}{k_0^4}\sin^4(\beta_s)(\varepsilon_1-\varepsilon_3)^2+4{\varepsilon_2}^2\varepsilon_3[\varepsilon_3-\frac{k^2_{I,II}(\beta_s)}{k_0^2}\sin^2(\beta_s)]\right]^{-1/2} \tag{6.101}$$

$$\Re = ERO \tag{6.102}$$

When the dipole is oriented in $\hat{z}$ direction, $\hat{u}=\hat{z}$, the components of the radiation fields in the Cartesian coordinate system are

$$E_{iI,II} = \left(\frac{-\omega\mu_0}{4\pi\ k_0^4}\right)\frac{e^{ik_{I,II}r\cos(\beta_s-\Theta)}}{r}(\Re)(A_{i3})_{\beta=\beta_s}\quad i=1,2,3\ or\ x,y,z \tag{6.103}$$

The coefficients A_{ij}, $(i,j)=1,2,3$ are evaluated at $k_z=k_{zI},\ k_{zII}$ and defined by (3.62) and (3.63). When the dipole is oriented in $\hat{y}$ direction, $\hat{u}=\hat{y}$, the components of the radiation fields in the Cartesian coordinate system are

$$E_{iI,II} = \left(\frac{-\omega\mu_0}{4\pi\ k_0^4}\right)\frac{e^{ik_{I,II}r\cos(\beta_s-\Theta)}}{r}(\Re)(A_{i2})_{\beta=\beta_s},\quad i=1,2,3\ or\ \ x,y,z \tag{6.104}$$

When the dipole is oriented in $\hat{x}$ direction, $\hat{u} = \hat{x}$, the components of the radiation fields in the Cartesian coordinate system are

$$E_{iI,II} = \left(\frac{-\omega\mu_0}{4\pi\ k_0^4}\right)\frac{e^{ik_{I,II}r\cos(\beta_s-\Theta)}}{r}(\Re)(A_{i1})_{\beta=\beta_s},\ i = 1,2,3\ or\ x,y,z \qquad (6.105)$$

The analytical results for the radiation fields given by (6.103)–(6.105) are compared with Bunkin's results [7]. Radiation fields given in [7] by (5.2) agrees with the results given by (6.103)–(6.105). In [7], the tensorial Green's function is represented using index notation and differential operators are introduced and used to solve the radiation problem. The derivation of the DGF presented here uses direct notation and dyadic Green's function is expressed as a sum of two single dyads in the spectral domain. Hence, in our case, a straightforward use of DGF with (9) leads to solution for the radiation fields. Besides, there are no numerical results given in [7].

In radiation problems, E_θ and E_ϕ components are commonly used to plot the radiation patterns. The expressions for E_θ and E_ϕ in terms of the electric field components E_x, E_y, and E_z are

$$E_\theta = E_x \cos\theta\cos\phi + E_y\cos\theta\sin\phi - E_z\sin\theta \qquad (6.106)$$

$$E_\phi = -E_x\sin\phi + E_y\cos\phi \qquad (6.107)$$

6.3 Numerical Results

This section consists of two sub-sections. Sect. 6.3.1 validates the results numerically with the existing results. Sect. 6.3.2 presents the radiation patterns in different frequency bands. In Sect. 6.3.2, we analyze the radiation fields derived in the previous section using the CMA diagram given in Fig. 6.12. The details of the CMA diagram which represents the wave propagation characteristics are given in Chap. 3. Both characteristic waves propagate in Regions 1, 3, 6 and 7.

6.3.1 Numerical Verification

The radiation patterns for the $\overline{E}_\phi$ component of the electric field when the dipole is oriented in x-direction and when $\frac{\omega_b}{\omega_p} = 1.8$ and $\frac{\omega}{\omega_p} = 2.4$ are given by Wu [8] on the x–z plane and the y–z plane. Our numerical results are plotted and compared with

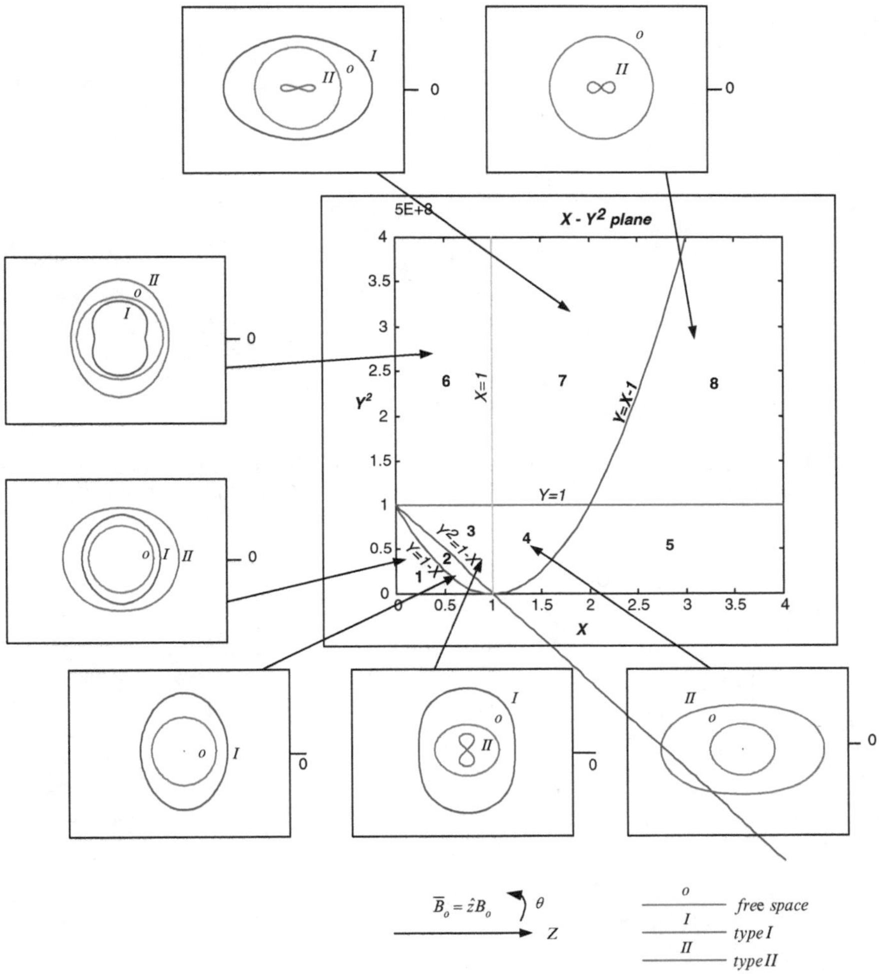

Fig. 6.12 CMA diagram to represent wave propagation characteristics

Wu's results in Figs. 6.13 and 6.14 below. The agreement is excellent. The given frequency of operation corresponds to region 1 in CMA diagram. In this region both type I and type II waves propagate.

The maximum radiation occurs at $\theta = 0°$ for type I wave and $\theta = 90°$ for type II wave when the observation point is on the $\phi = 90°$ plane (y–z plane), respectively. On this plane of observation, type I wave does not radiate in the $\theta = 90°$ direction. On the other hand, type II wave radiates in all directions. When the observation point is on the $\phi = 0°$ plane (x–z plane), the maximum and minimum radiation directions remain same for type I wave. However, the maximum direction of radiation changes from $\theta = 0°$ to $\theta = 90°$ for type II wave. Type II wave again radiates in all directions on this plane.

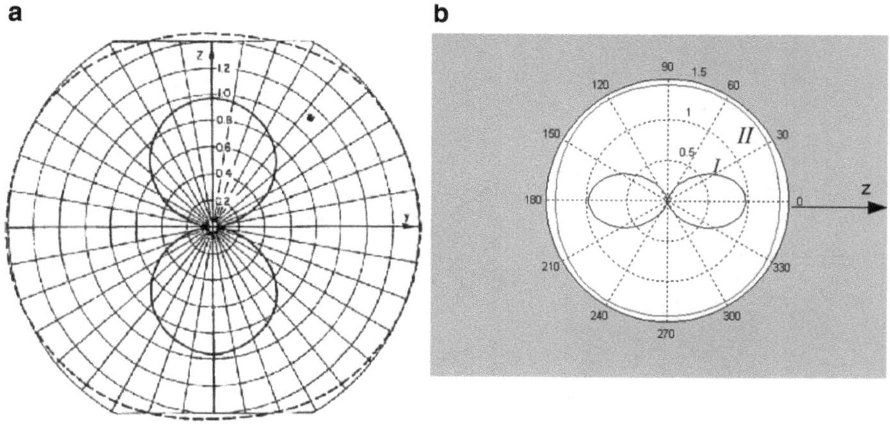

Fig. 6.13 (**a**) x-directed dipole E_ϕ component on the y–z plane, Wu's result. (**b**) x-directed dipole E_ϕ component on the y–z plane, our result $X = 0.1736, Y = 0.75,$ Region 1 in CMA diagram

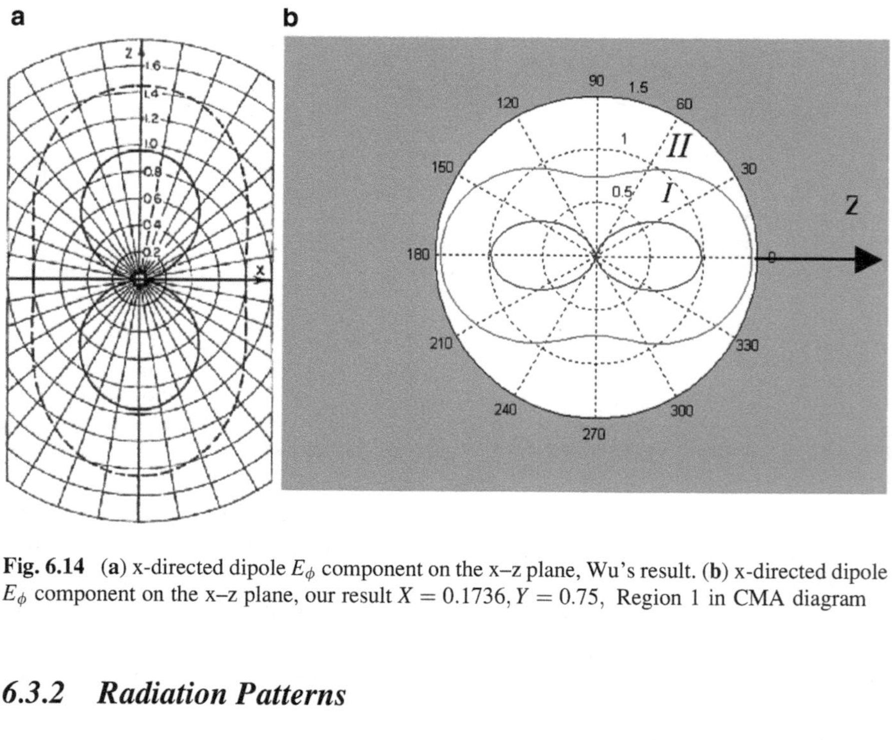

Fig. 6.14 (**a**) x-directed dipole E_ϕ component on the x–z plane, Wu's result. (**b**) x-directed dipole E_ϕ component on the x–z plane, our result $X = 0.1736, Y = 0.75,$ Region 1 in CMA diagram

6.3.2 Radiation Patterns

In this section, the radiation patterns are obtained for $\hat{\theta}$ and $\hat{\phi}$ components of the radiation field on the $\phi = 0°$ plane for both type I and type II waves when the dipole is oriented in the z-direction and in the x-direction. We analyzed the radiation patterns in three different regions of the CMA diagram. The regions are chosen such that there exist wave propagation due to both type I and type II waves, due to only type I wave and due to only type II wave. As described in Chap. 3, in Region 1 of the CMA diagram both type I and type II waves propagate, in Region 2 only type I wave propagates and in Region 4 only type II wave propagates. Hence, Regions 1, 2 and 4 of the CMA diagram are the regions for which we will investigate the radiation characteristics of the waves.

6.3.2.1 Radiation in Region 1

The wave normal surface in this region is given in Fig. 6.15. This region corresponds to a high frequency region. Based on the wave normal surface given in the above figure, in this region both type I and type II waves propagate. Hence we expect both waves to radiate in this region when the frequency of the operation falls within the specified range above. The radiation pattern for the $\hat{\theta}$ component and the $\hat{\phi}$ component of the $\overline{E}$ field for z-directed dipole on the x–z plane are shown in Fig. 6.16a, b, respectively.

When the dipole is z-directed, the $\hat{\theta}$ component of the radiation field for type I wave has the maximum radiation at $\theta = 54.9°$. It has one symmetrical major lobe with respect to $\theta = 90°$. The minimum radiation occurs at $\theta = 0°$ for the type I wave. Type II wave has two symmetrical lobes with respect to $\theta = 90°$. The maximum radiation happens at $\theta = 39.6°$. The minimum radiation is at $\theta = 0°$, $90°$ for the type II wave. When we compare maximum radiation fields for both types of waves, the ratio of their magnitudes is $\frac{|E_{\theta I}|}{|E_{\theta II}|} = 2.96$. The $\hat{\phi}$ component of the radiation field for type I wave has the maximum radiation at $\theta = 39.6°$. It has two symmetrical lobes with respect to $\theta = 90°$. The minimum radiation is at $\theta = 0°$, $90°$ for the type I wave. The type II wave has also two symmetrical lobes with respect to $\theta = 90°$. It has maximum radiation at $\theta = 47.1°$. The minimum radiation happens at $\theta = 0°$, $90°$. When we compare maximum radiation fields for both types of waves, the ratio of their magnitudes is $\frac{|E_{\phi I}|}{|E_{\phi II}|} = 1.79$.

When the dipole is oriented in the x-direction, the radiation patterns for the $\hat{\theta}$ component and the $\hat{\phi}$ component of the $\overline{E}$ field are shown in Fig. 6.17. The $\hat{\theta}$

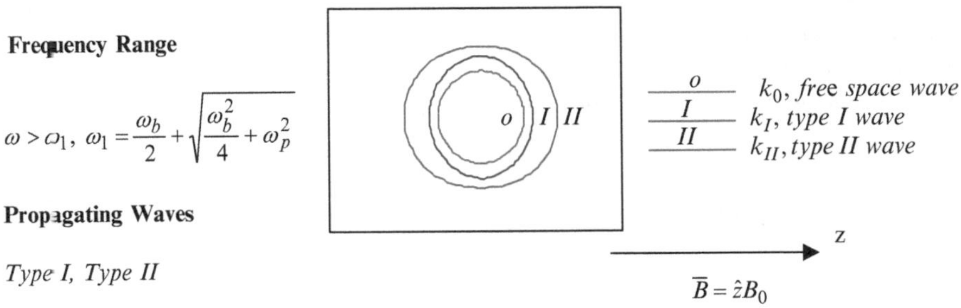

Fig. 6.15 Wave normal surface in Region 1 when $X = 0.44, Y = 0.37$

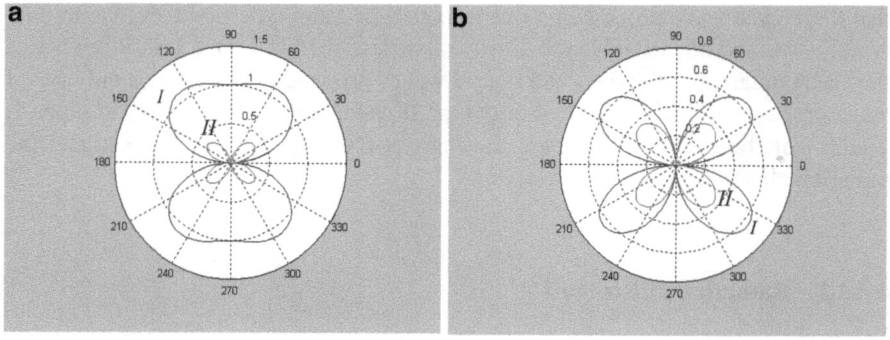

Fig. 6.16 (**a**) E_θ component of the radiation field for z-directed dipole on the on x–z plane. (**b**) E_ϕ component of the radiation field for z-directed dipole on the x–z plane

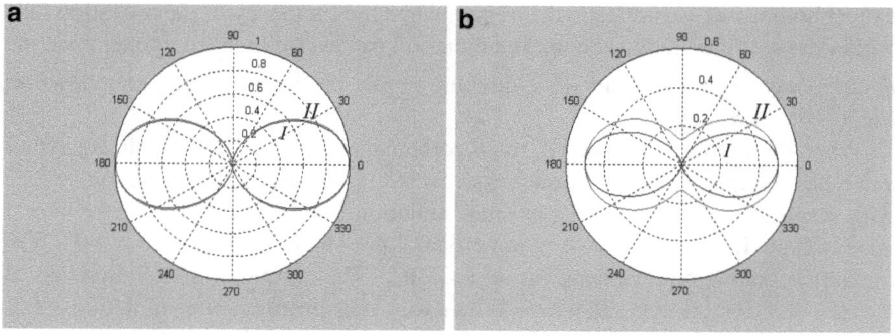

Fig. 6.17 (**a**) E_θ component of the radiation field for x-directed dipole on the x–z plane. (**b**) E_ϕ component of the radiation field for x-directed dipole on the x–z plane

component of the radiation field for both type I and type II waves have the maximum radiation at $\theta = 0°$. They have symmetrical radiation pattern with respect to $\theta = 90°$. The minimum radiation occurs at $\theta = 90°$ for both the types of waves. Along the direction of maximum radiation for both types of waves, the ratio of their magnitudes is $\frac{|E_{\theta I}|}{|E_{\theta II}|} = 1.02$. The $\hat{\phi}$ component of the radiation field for both type I and type II waves have the maximum radiation at $\theta = 0°$. They have symmetrical radiation pattern with respect to $\theta = 90°$. The minimum radiation for both waves are at $\theta = 90°$. While the type I wave does not radiate at $\theta = 90°$, the type II wave radiates in all directions. When there is a maximum radiation for both types of waves, the ratio of their magnitudes is $\frac{|E_{\phi I}|}{|E_{\phi II}|} = 1.01$.

6.3.2.2 Radiation in Region 2

The wave normal surface in this region is given in Fig. 6.18.

In Region 2, only type I waves can propagate. Hence, we expect only radiation due to type I wave in this region. The numerical plots for the radiation patterns

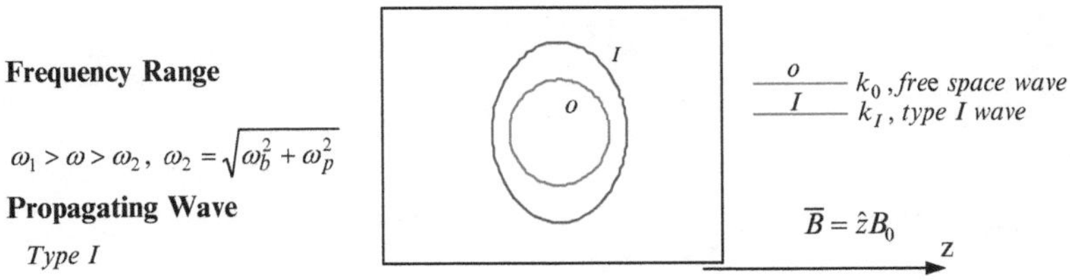

Fig. 6.18 Wave normal surface in Region 2 when $X = 0.6083$ and $Y = 0.4386$

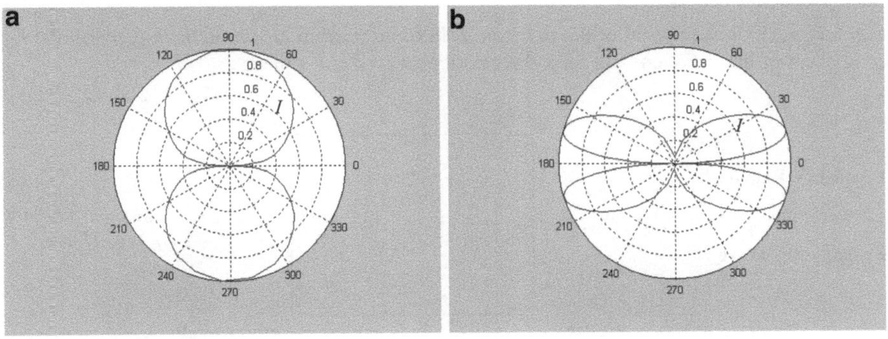

Fig. 6.19 (**a**) E_θ component of the radiation field for z-directed dipole on the x–z plane. (**b**) E_ϕ component of the radiation field for z-directed dipole on the x–z plane

shown in Figs. 6.19 and 6.20 confirm this. The radiation pattern for the $\hat{\theta}$ component and the $\hat{\phi}$ component of the $\overline{E}$ field for z-directed dipole on x–z plane are shown in Fig. 6.19a, b, respectively. When the dipole is z-directed, the $\hat{\theta}$ component of the radiation field for type I wave has the maximum radiation at $\theta = 90°$. It has one major lobe and has no radiation at $\theta = 0°$. The radiation pattern looks like the radiation pattern of an antenna in the uniaxial plasma. The ϕ component of type I wave has symmetrical lobes with respect to $\theta = 90°$. The maximum radiation occurs at $\theta = 17.1°$. The minimum radiation for type I wave is at $\theta = 0°,\ 90°$.

When the dipole is oriented in the x-direction, the radiation patterns for the $\hat{\theta}$ component and the $\hat{\phi}$ component of the $\overline{E}$ field are shown in Fig. 6.20. The $\hat{\theta}$ and $\hat{\phi}$ components of the radiation field for type I wave has the maximum radiation at $\theta = 0°$. The minimum radiation occurs at $\theta = 90°$ for both components. The radiation patterns have narrow beamwidth and hence the antenna is very directive in this region for $\hat{\theta}$ and $\hat{\phi}$ components of the radiation field for type I wave.

6.3.2.3 Radiation in Region 4

The wave normal surface in this region is given in Fig. 6.21. In Region 4, only type II wave can propagate. So the radiation should be due to only type II wave in this region.

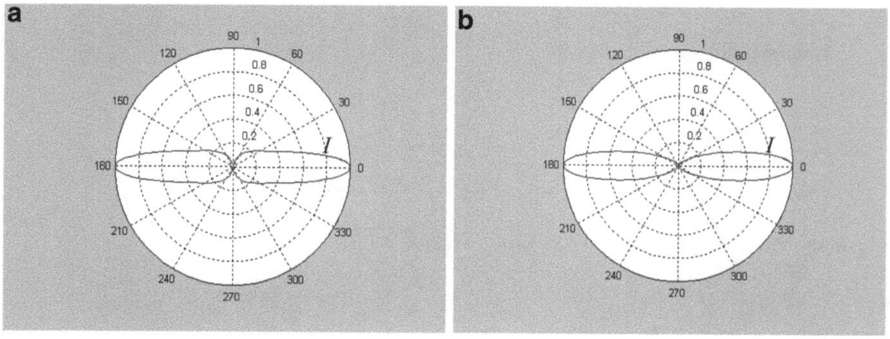

Fig. 6.20 (**a**) E_θ component of the radiation field for x-directed dipole on the x–z plane. (**b**) E_ϕ component of the radiation field for x-directed dipole on the x–z plane

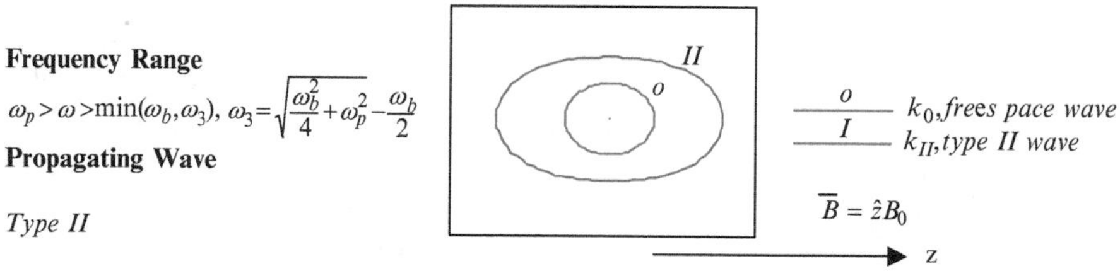

Fig. 6.21 Wave normal surface in Region 4 when $X = 1.5041$ and $Y = 0.6897$

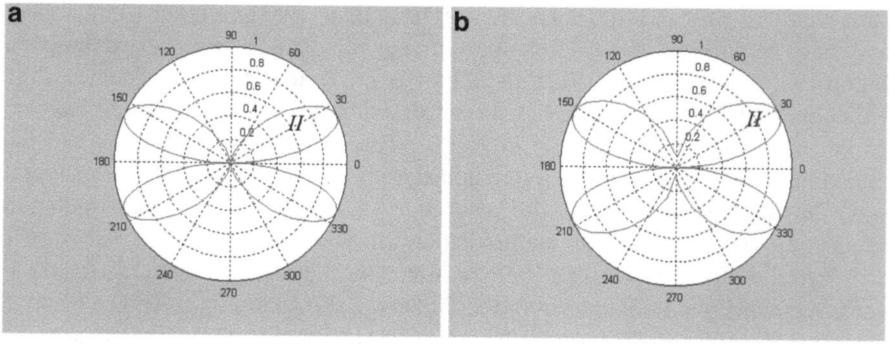

Fig. 6.22 (**a**) E_θ component of the radiation field for z-directed dipole on the x–z plane. (**b**) E_ϕ component of the radiation field for z-directed dipole on the x–z plane

The numerical plots for the radiation patterns shown in Figs. 6.22 and 6.23 confirm this. The radiation pattern for the $\hat{\theta}$ component and the $\hat{\phi}$ component of the $\bar{E}$ field for the z-directed dipole on x–z plane are shown in Fig. 6.22a, b, respectively. When the dipole is z-directed, the $\hat{\theta}$ component of the radiation field for type II wave has the maximum radiation at $\theta = 24.1°$. The $\hat{\phi}$ component of type II wave has the maximum radiation at $\theta = 27.5°$. The minimum radiation is at $\theta = 0°,\ 90°$ for both waves.

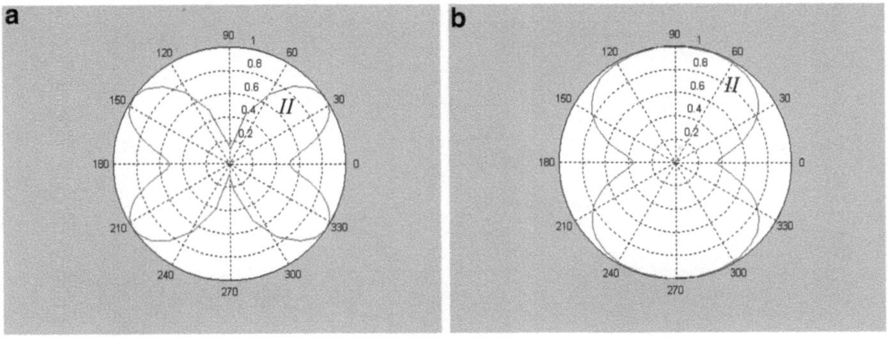

Fig. 6.23 (**a**) E_θ component of the radiation field for x-directed dipole on the x–z plane. (**b**) E_ϕ component of the radiation field for x-directed di pole on the x–z plane

When the dipole is oriented in the x-direction, the radiation patterns for the $\hat{\theta}$ component and the $\hat{\phi}$ component of the $\overline{E}$ field are shown in Fig. 6.23. The $\hat{\theta}$ component of the radiation field for type II wave has the maximum radiation at $\theta = 31.6°$. It has the symmetrical radiation pattern with respect to $\theta = 90°$. The $\hat{\phi}$ component of type II wave has one major lobe which has the maximum radiation occurs at $\theta = 90°$. The minimum radiation occurs at $\theta = 90°$ and $\theta = 0°$ for the $\hat{\theta}$ component and at $\theta = 0°$ only for the $\hat{\phi}$ component of the radiation field for the type II wave. The antenna radiates in all directions for both components for this case.

References

1. C.M. Bender and S.A. Orszag, *Advanced Mathematical Methods for Scientists and Engineers*, New York, McGraw-Hill, 1978, pp. 280–302.
2. J.A. Kong, *Theory of Electromagnetic Waves*, New York, Wiley, 1975, pp. 59–62, 205–212.
3. W.C. Chew, *Waves and Fields in Inhomogeneous Media*, IEEE Press, 1995.
4. R. Courant and D. Hilbert, *Methods of Mathematical Physics*, Vol. 1, Interscience Publisers, New York, Second Printing, 1955.
5. M. Abramowitz and I. Stegun, *Handbook of Mathematical Functions*, Dover Publications, New York, 1972, Chapter 9, pp. 361.
6. L.B. Felsen and N. Marcuvitz, *Radiation and Scattering of Waves*, Prentice-Hall, Englewood Cliffs, NJ, 1973.
7. F.V. Bunkin, "On radiation in anisotropic media," *Sov. Phys. JETP, Engl. Transl.*, 5, 277–295, 1957.
8. C.P. Wu, "Radiation from dipoles in an magneto-ionic medium," *IEEE Trans., Antennas Propag.*, AP-11, pp. 681–689, 1963.

Chapter 7
Wave Theory of Composite Layered Structures

The advances in microwave communication systems require material properties which can give optimum device performance. New materials are being developed in order to meet the requirements of design processes as technology matures. These new materials are composite mixtures of isotropic, anisotropic or gyrotropic materials. The purpose of such materials is to obtain more desirable constitutive parameters through a design process that can predict the effective properties of such a mixture. One such benefit, besides having the permittivity or permeability that exactly fits the application, is to use materials with lower cost to create a composite material that might otherwise be found in much more expensive bulk form. As a consequence, the analysis of the wave propagation through multilayered composite structures carries great importance in the development of the microwave devices with the desired material properties. If the layered structure is considered to be consisted of layered general anisotropic media, then composite mixture can be obtained by setting the desired arbitrary layer to be isotropic, electrically, magnetically gyrotropic, uniaxially or biaxially anisotropic material by changing the elements of permittivity and permeability tensors for the layered structure. In the following sections wave propagation through multilayered composite structure will be given using step by step procedure. General anisotropic medium in the composite layered structure is defined with permittivity and permeability tensors, $\bar{\bar{\varepsilon}}$ and $\bar{\bar{\mu}}$, respectively.

7.1 Wave Propagation in Multilayered Isotropic Media

The fundamental concept of wave propagation through layered general anisotropic media can be better understood by first studying the wave propagation through multilayered isotropic media. The reflected and transmitted waves from layered isotropic media can be found using the wave matrix approach [1]. In Sect. 7.1.1, we will first investigate wave reflection and transmission from single-layered isotropic media. We will then extend this analysis to multilayered structure in Sect. 7.1.2.

A. Eroglu, *Wave Propagation and Radiation in Gyrotropic and Anisotropic Media*, DOI 10.1007/978-1-4419-6024-5_7, © Springer Science+Business Media, LLC 2010

7.1.1 Single-Layered Isotropic Media

Consider a plane wave which is incident from Region 0 with a plane of incidence parallel to the x–z plane as shown in Fig. 7.1. Plane of incidence is formed by the incident wave vector $\bar{k}$ and normal to the boundary surface. Incident wave can be a horizontally polarized wave ($E - wave$), or TE wave, which is perpendicular to the plane of incidence (POI) or vertically polarized wave ($H - wave$) which is parallel to POI. We will study two cases of wave polarization for the isotropic single-layered structure.

7.1.1.1 TE Wave

In the first case, we assume the incident wave is TE wave and linearly polarized in the y-direction. The electric field can be written as

$$\bar{E}_i = \hat{y}E_0 e^{i(k_{ix}x + k_{iz}z)} e^{-i\omega t} \tag{7.1}$$

$$\bar{E}_r = \hat{y}R^{TE}E_0 e^{i(-k_{rx}x + k_{rz}z)} e^{-i\omega t} \tag{7.2}$$

$$\bar{E}_t = \hat{y}T^{TE}E_0 e^{i(k_{tx}x + k_{tz}z)} e^{-i\omega t} \tag{7.3}$$

Magnetic field vector, $\bar{H}$, can be found using Faraday's law as

$$\frac{\partial \bar{H}}{\partial t} = -\frac{1}{\mu}\nabla \times \bar{E} \tag{7.4}$$

as

$$\bar{H}_i = (-\hat{x}k_{iz} + \hat{z}k_{ix})\left(\frac{E_0}{\omega\mu_0}\right) e^{i(k_{ix}x + k_{iz}z)} e^{-i\omega t} \tag{7.5}$$

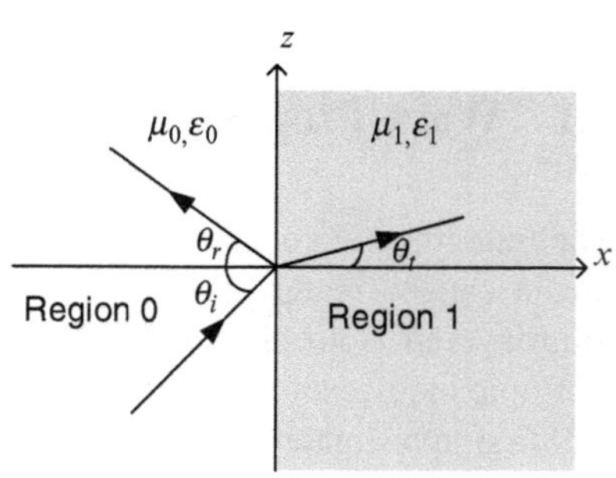

Fig. 7.1 Reflection and transmission of waves in isotropic single-layered structure

$$\bar{H}_r = -(\hat{x}k_{rz} + \hat{z}k_{rx})R^{TE}\left(\frac{E_0}{\omega\mu_0}\right)e^{i(-k_{rx}x+k_{rz}z)}e^{-i\omega t} \tag{7.6}$$

$$\bar{H}_t = (-\hat{x}k_{tz} + \hat{z}k_{tx})T^{TE}\left(\frac{E_0}{\omega\mu_1}\right)e^{i(k_{tx}x+k_{tz}z)}e^{-i\omega t} \tag{7.7}$$

In (7.1)–(7.7), R, and T are called reflection and transmission coefficients, respectively. The superscript refers the type of wave, i.e. TE or TM. Using the continuity of the tangential components of electric fields at the boundary where x = 0, we find

$$1 + R^{TE} = T^{TE} \tag{7.8}$$

$$\left(\frac{k_x}{\mu_0}\right)(1 - R^{TE}) = \left(\frac{k_{tx}}{\mu_1}\right)T^{TE} \tag{7.9}$$

with the phase matching condition

$$k_{iz} = k_{rz} = k_{tz} \tag{7.10}$$

Please note the wave vectors in Region 0 and Region 1 are

$$\bar{k}_i = \hat{x}k_{ix} + \hat{z}k_{iz} \tag{7.11}$$

$$\bar{k}_r = -\hat{x}k_{rx} + \hat{z}k_{rz} \tag{7.12}$$

$$\bar{k}_t = \hat{x}k_{tx} + \hat{z}k_{tz} \tag{7.13}$$

and corresponding dispersion relations are obtained as

$$k_i^2 = k_{ix}^2 + k_{iz}^2 = \omega^2\mu_0\varepsilon_0 \tag{7.14}$$

$$k_r^2 = k_{rx}^2 + k_{rz}^2 = \omega^2\mu_0\varepsilon_0 \tag{7.15}$$

$$k_t^2 = k_{tx}^2 + k_{tz}^2 = \omega^2\mu_1\varepsilon_1 \tag{7.16}$$

Solving (7.8) and (7.9) for R^{TE} and T^{TE}, we find

$$R^{TE} = \frac{1 - \left(\frac{\mu_0 k_{tx}}{\mu_1 k_{ix}}\right)}{1 + \left(\frac{\mu_0 k_{tx}}{\mu_1 k_{ix}}\right)} \tag{7.17}$$

$$T^{TE} = \frac{2}{1 + \left(\frac{\mu_0 k_{tx}}{\mu_1 k_{ix}}\right)} \tag{7.18}$$

R^{TE} and T^{TE} are called Fresnel reflection and transmission coefficients, respectively for TE waves. The relation between angle of reflection and transmission are obtained using the phase matching condition given in (7.10), (7.14), (7.15) as

$$k_i \sin\theta_i = k_r \sin\theta_r = k_t \sin\theta_t \tag{7.19}$$

$$k_i = k_r \tag{7.20}$$

(7.19) and (7.20) leads to

$$\theta_i = \theta_r \tag{7.21}$$

$$\frac{\sin\theta_t}{\sin\theta_i} = \frac{\sqrt{\mu_0\varepsilon_0}}{\sqrt{\mu_1\varepsilon_1}} \tag{7.22}$$

7.1.1.2 TM Wave

Now consider, incident wave is TM wave for the structure shown in Fig. 7.1. Assume the incident magnetic wave is linearly polarized in y-direction. Then, the magnetic field wave can be written as

$$\bar{H}_i = \hat{y}H_0 e^{i(k_{ix}x + k_{iz}z)} e^{-i\omega t} \tag{7.23}$$

$$\bar{H}_r = \hat{y}R^{TM} H_0 e^{i(-k_{rx}x + k_{rz}z)} e^{-i\omega t} \tag{7.24}$$

$$\bar{H}_t = \hat{y}T^{TM} H_0 e^{i(k_{tx}x + k_{tz}z)} e^{-i\omega t} \tag{7.25}$$

Electric filed vector, $\bar{E}$, can be found using Ampere's law

$$\frac{\partial \bar{E}}{\partial t} = \frac{1}{\varepsilon}\nabla \times \bar{H} \tag{7.26}$$

as

$$\bar{E}_i = (\hat{x}k_{iz} - \hat{z}k_{ix})\left(\frac{H_0}{\omega\varepsilon_0}\right) e^{i(k_{ix}x + k_{iz}z)} e^{-i\omega t} \tag{7.27}$$

$$\bar{E}_r = (\hat{x}k_{rz} + \hat{z}k_{rx})R^{TM}\left(\frac{H_0}{\omega\varepsilon_0}\right) e^{i(-k_{rx}x + k_{rz}z)} e^{-i\omega t} \tag{7.28}$$

$$\bar{E}_t = (\hat{x}k_{tz} - \hat{z}k_{tx})T^{TM}\left(\frac{H_0}{\omega\varepsilon_1}\right) e^{i(k_{tx}x + k_{tz}z)} e^{-i\omega t} \tag{7.29}$$

The Fresnel coefficients for TM waves are similarly calculated using boundary conditions when x = 0 as

$$R^{TM} = \frac{1 - \left(\frac{\varepsilon_0 k_{tx}}{\varepsilon_1 k_{ix}}\right)}{1 + \left(\frac{\varepsilon_0 k_{tx}}{\varepsilon_1 k_{ix}}\right)} \tag{7.30}$$

$$T^{TE} = \frac{2}{1 + \left(\frac{\varepsilon_0 k_{tx}}{\varepsilon_1 k_{ix}}\right)} \tag{7.31}$$

When (7.17), (7.18) and (7.26), (7.27) are compared, it is seen that Fresnel coefficients of TM waves and TE waves can be obtained from each other with the application of the duality following relations

$$\bar{E} \to \bar{H}, \quad \bar{H} \to -\bar{E}, \quad \mu \leftrightarrow \varepsilon \tag{7.32}$$

7.1.2 Multilayered Isotropic Media

In this section, the analysis of wave propagation for single-layered isotropic media described in Sect. 7.2.1 will be extended to multilayered isotropic media using wave matrices [1]. Consider the multilayered isotropic structure illustrated in Fig. 7.2 which has its stratification in z-direction. Assume plane wave which has POI parallel to x–z plane is incident from Region 0. The layered medium has t = n + 1 isotropic layer with μ_l permeability and ε_l permittivity where $l = 0, 1, 2 \ldots t$.

7.1.2.1 TE Wave

Initially, we analyze the wave propagation for TE waves. The incident electric field vector is linearly polarized in the y-direction and given as

$$\bar{E}_i = \hat{y} E_0 e^{i(k_{ix}x - k_{iz}z)} e^{-i\omega t} \tag{7.33}$$

Then, the total electric field in Region l can be written as

$$\bar{E}_l = \hat{y}\left(A_l e^{ik_{lz}z} + B_l e^{-ik_{lz}z}\right) e^{ik_x x} e^{-i\omega t} \tag{7.34}$$

In (7.34), A_l is the amplitude of the all wave components traveling in (+) z-direction whereas B_l is the amplitude of the all wave components traveling in

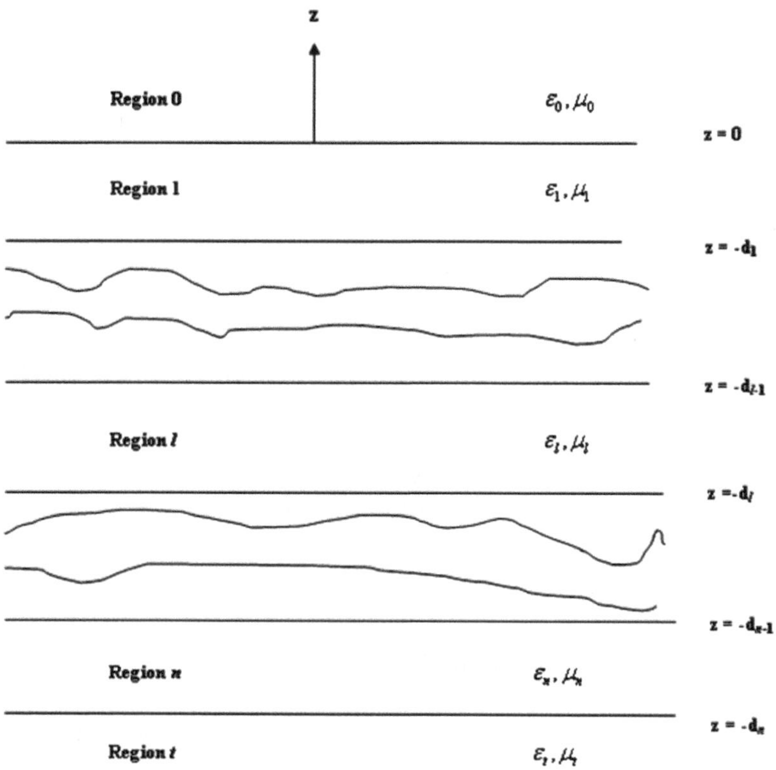

Fig. 7.2 Reflection and transmission of waves in isotropic multilayered structure

(−) z-direction. The corresponding magnetic field vectors in the same region are found from (7.4) as

$$H_l = \left(\left(-\hat{x}\frac{k_{lz}}{\omega\mu_l} + \hat{z}\frac{k_x}{\omega\mu_l}\right)A_l e^{ik_{lz}z} + \left(\hat{x}\frac{k_{lz}}{\omega\mu_l} + \hat{z}\frac{k_x}{\omega\mu_l}\right)B_l e^{ik_{lz}z}\right)e^{ik_x x}e^{-i\omega t} \quad (7.35)$$

Application of boundary conditions for n + 1 boundaries gives (2n + 2) equations. Since there are total of (2n + 2) unknowns, they can be determined using matrix solution. Using (7.34) and (7.35) and continuity of the tangential components of the field vectors along the boundary at $z = -d_l$, we relate the wave components

$$A_l e^{-ik_{lz}d_l} + B_l e^{ik_{lz}d_l} = A_{l+1}e^{-ik_{(l+1)z}d_l} + B_{l+1}e^{ik_{(l+1)z}d_l} \quad (7.36)$$

$$\frac{\mu_{(l+1)}k_{lz}}{\mu_l k_{(l+1)z}}\left[A_l e^{-ik_{lz}d_l} - B_l e^{ik_{lz}d_l}\right] = \left[A_{l+1}e^{-ik_{(l+1)z}d_l} - B_{l+1}e^{ik_{(l+1)z}d_l}\right] \quad (7.37)$$

Equations (7.36) and (7.37) can be put in the following form

$$A_{l+1}e^{-ik_{(l+1)z}d_l} = \frac{1}{2}\left(1+\frac{\mu_{(l+1)}k_{lz}}{\mu_l k_{(l+1)z}}\right)\left(A_l e^{-ik_{lz}d_l} + R_{(l+1)}B_l e^{ik_{lz}d_l}\right) \tag{7.38}$$

$$B_{l+1}e^{ik_{(l+1)z}d_l} = \frac{1}{2}\left(1+\frac{\mu_{(l+1)}k_{lz}}{\mu_l k_{(l+1)z}}\right)\left(R_{(l+1)}A_l e^{-ik_{lz}d_l} + B_l e^{ik_{lz}d_l}\right) \tag{7.39}$$

where

$$R_{(l+1)l} = -R_{l(l+1)} = \frac{\left(1-\Gamma^e_{(l+1)l}\right)}{\left(1+\Gamma^e_{(l+1)l}\right)} \tag{7.40}$$

and

$$\Gamma^e_{(l+1)l} = \frac{\mu_{(l+1)}k_{lz}}{\mu_l k_{(l+1)z}} \tag{7.41}$$

$\Gamma^e_{(l+1)l}$ is called the reflection coefficient at the boundary for TE waves between regions l and $l+1$. The first subscript, shows that the incident wave region. Equations (7.38) and (7.39) can now be put in matrix form as

$$\begin{pmatrix} A_{l+1}e^{-ik_{(l+1)z}d_{l+1}} \\ B_{l+1}e^{ik_{(l+1)z}d_{l+1}} \end{pmatrix} = F_{(l+1)l}\begin{pmatrix} A_l e^{-ik_{lz}d_l} \\ B_l e^{ik_{lz}d_l} \end{pmatrix} \tag{7.42}$$

$F_{(l+1)l}$ is the forward propagating matrix and defined as

$$F_{(l+1)l} = \frac{1}{2}\left(1+\Gamma^e_{(l+1)l}\right)\begin{pmatrix} e^{-ik_{(l+1)z}(d_{l+1}-d_l)} & R_{(l+1)l}e^{-ik_{(l+1)z}(d_{l+1}-d_l)} \\ R_{(l+1)l}e^{ik_{(l+1)z}(d_{l+1}-d_l)} & e^{ik_{(l+1)z}(d_{l+1}-d_l)} \end{pmatrix} \tag{7.43}$$

When the wave is in Region t = n + 1, there is only transmission coefficient, T, because this is a semi infinite region and there are no reflected wave components. Transmission coefficient, T, is found from (7.42) as by setting $A_{l+1} = 0$, and $B_{l+1} = T$ as

$$\begin{pmatrix} 0 \\ T \end{pmatrix} = F_{tn}\begin{pmatrix} A_l e^{-ik_{lz}d_l} \\ B_l e^{ik_{lz}d_l} \end{pmatrix} \tag{7.44}$$

where

$$F_{tn} = \frac{1}{2}\left(1+\Gamma^e_{tn}\right)\begin{pmatrix} e^{ik_{tz}d_n} & R_{tn}e^{ik_{tz}d_n} \\ R_{tn}e^{-ik_{tz}d_n} & e^{-ik_{tz}d_n} \end{pmatrix} \tag{7.45}$$

7.1.2.2 TM Wave

The Fresnel coefficients for TM waves are simply obtained by application of duality relation given in (7.32). So,

$$\Gamma^{e}_{(l+1)l} = \frac{\mu_{(l+1)}k_{lz}}{\mu_{l}k_{(l+1)z}} \rightarrow \Gamma^{m}_{(l+1)l} = \frac{\varepsilon_{(l+1)}k_{lz}}{\varepsilon_{l}k_{(l+1)z}} \tag{7.46}$$

We can also represent the wave amplitudes in any region using the wave amplitudes in other regions through wave propagation matrices in (7.43). The transmission and reflection coefficients are determined from the elements of the forward propagating matrix.

7.2 Wave Propagation in Multilayered Anisotropic Media

In this section, we will investigate the wave propagation through multilayered anisotropic media using the theory outlined in Sect. 7.1. The analysis will be first described using single, two-layered and then multilayered uniaxially anisotropic media in Sect. 7.2.1–7.2.5. The analysis will then be extended to multilayered general layered anisotropic media in Sect. 7.2.5.

7.2.1 Single-Layered Anisotropic Media: Vertically Uniaxial Case

In this section, the analysis of the wave propagation for a single layer anisotropic media is described. The anisotropic layer is unaxially anisotropic medium with its optic axis oriented in z-direction (vertically uniaxially case). The wave is incident from isotropic region which is designated by Region 0. The incident wave in isotropic region can be either TE wave or horizontally polarized or TM wave or vertically polarized as described in Sect. 7.1.2. The transmitted wave in anisotropic region is an $o-ordinarywave$ if the incident wave is TE wave and $e-ordinarywave$ if the incident wave is TM wave as described in Chap. 2. The reflected wave in isotropic region is horizontally polarized wave when incident wave is TE wave and vertically polarized wave when incident wave is TM wave. In this section, we will introduce two unit vectors, $\hat{h}$ and $\hat{v}$, to represent horizontally polarized (TE) and vertically polarized waves (TM), respectively [2]. The unit vectors, $\hat{h}$ and $\hat{v}$, and the wave vectors illustrated in Fig. 7.3 are defined as

$$\hat{h}_0(\pm k_{0z}) = \frac{\hat{z}\times\bar{k}_0}{k_\rho} = \frac{\left(-\hat{x}k_y + \hat{y}k_x\right)}{k_\rho} \tag{7.47}$$

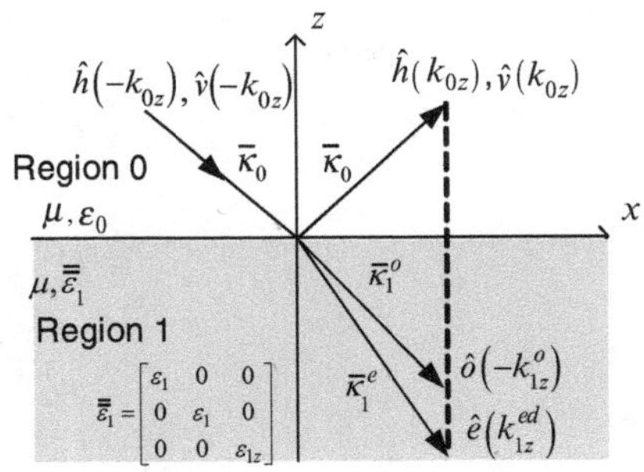

Fig. 7.3 Reflection and transmission of waves in single-layered anisotropic slab

$$\hat{v}_0(k_{0z}) = \frac{\hat{h}_0(k_{0z}) \times \bar{k}_0}{k_0} = \frac{1}{k_0}\left[\frac{k_{oz}(\hat{x}k_x + \hat{y}k_y)}{k_\rho} - \hat{z}k_\rho\right] \tag{7.48}$$

$$\hat{v}_0(-k_{0z}) = \frac{\hat{h}_0(-k_{0z}) \times \bar{\kappa}_0}{k_0} = \frac{1}{k_0}\left[-\frac{k_{oz}(\hat{x}k_x + \hat{y}k_y)}{k_\rho} - \hat{z}k_\rho\right] \tag{7.49}$$

$$\hat{o}(-k_{1z}^o) = \frac{\hat{z} \times \bar{\kappa}_1^o}{\left|\hat{z} \times \bar{\kappa}_1^o\right|} = \frac{(-\hat{x}k_y + \hat{y}k_x)}{k_\rho} = \hat{h}_1(-k_{1z}) \tag{7.50}$$

$$\hat{e}(k_{1z}^{ed}) = \frac{\hat{h}_1(-k_{1z}) \times \bar{\kappa}_{1u}}{k_{1u}} = \frac{1}{k_{1u}}\left[\frac{-\bar{k}_\rho k_{1z}^{ed}\varepsilon_{1z}}{\varepsilon_1 k_\rho} - \hat{z}k_\rho\right] \tag{7.51}$$

$$\bar{k}_0 = \bar{k}_\rho + \hat{z}k_0 \tag{7.52}$$

$$\bar{\kappa}_0 = \bar{k}_\rho - \hat{z}k_0 \tag{7.53}$$

$$\bar{k}_\rho = \hat{x}\bar{k}_x + \hat{y}k_y \tag{7.54}$$

$$\bar{k}_1^o = \bar{k}_\rho + \hat{z}k_{1z}^o \tag{7.55}$$

$$\bar{\kappa}_1^o = \bar{k}_\rho - k_{1z}^\circ \hat{z} \tag{7.56}$$

$$\bar{k}_1^e = \bar{k}_\rho + \hat{z}k_{1z}^{eu} \tag{7.57}$$

$$\bar{\kappa}_1^e = \bar{k}_\rho + k_{1z}^{ed}\hat{z} \tag{7.58}$$

$$\bar{k}_{1u} = \bar{\bar{\varepsilon}} \cdot \bar{k}_1^e \tag{7.59}$$

$$\bar{\kappa}_{1u} = \bar{\bar{\varepsilon}} \cdot \bar{\kappa}_1^e \tag{7.60}$$

The wave numbers are found from the dispersion relations as

$$k_{0z}^{\circ} = \left[k_0^2 - k_\rho^2\right]^{\frac{1}{2}} \tag{7.61}$$

$$k_{1z}^{\circ} = \left[k_1^2 - k_\rho^2\right]^{\frac{1}{2}} \tag{7.62}$$

$$k_{0,1} = \omega\sqrt{\mu\varepsilon_{0,1}} \tag{7.63}$$

$$\begin{bmatrix} k_{1z}^{eu} \\ k_{1z}^{ed} \end{bmatrix} = \pm\frac{1}{\varepsilon_{1z}}\left[k_1^2\varepsilon_{1z}^2 - \varepsilon_1\varepsilon_{1z}k_\rho^2\right]^{1/2} \tag{7.64}$$

7.2.1.1 TE Wave

Assume the incident wave from Region 0 is TE wave and given by

$$\bar{E}_{i,0} = \hat{h}_0(-k_{0z})e^{i(\bar{\kappa}_o\cdot\bar{r})}e^{-i\omega t} \tag{7.65}$$

The first subscript shows the wave type (incident, reflected, or transmitted in Region 1).The reflected wave is

$$\bar{E}_{r,0} = \left[\hat{h}_0(k_{0z})R_{01HH}e^{i(\bar{k}_o\cdot\bar{r})}\right]e^{-i\omega t} \tag{7.66}$$

R_{01HH} is the reflection coefficient when incident and reflected waves are both horizontally polarized. The transmitted wave in Region 1 can be expressed as

$$\bar{E}_{t,1} = \left[\hat{o}(-k_{1z}^o)T_{Ho}e^{i(\bar{\kappa}_1^o\cdot\bar{r})}\right]e^{-i\omega t} \tag{7.67}$$

T_{Ho} is the transmission coefficient when incident wave is horizontally polarized and transmitted wave is an ordinary wave The reflection and transmission coefficients, Fresnel coefficients, are found using boundary conditions at $z = 0$ as described in the previous section. Application of boundary conditions $z = 0$ gives following equations

$$\hat{z}\times\bar{E}_0 = \hat{z}\times\bar{E}_1 \tag{7.68}$$

$$\hat{z}\times\nabla\times\bar{E}_0 = \hat{z}\times\nabla\times\bar{E}_1 \tag{7.69}$$

where

$$\bar{E}_0 = \bar{E}_{i,0} + \bar{E}_{r,0} \tag{7.70}$$

$$\bar{E}_1 = \bar{E}_{t,1} \tag{7.71}$$

Substituting (7.65)–(7.67), and (7.70), (7.71) into (7.68) and (7.69) gives

$$R_{01HH} = \frac{k_{oz} - k_{1z}^o}{k_{oz} + k_{1z}^o} \tag{7.72}$$

$$T_{Ho} = \frac{2k_{oz}}{\left(k_{oz} + k_{1z}^o\right)} \tag{7.73}$$

7.2.1.2 TM Wave

Now, assume the incident wave from Region 0 is TM wave and given by

$$\bar{E}_{i,0} = \hat{v}_0(-k_{0z})e^{i(\bar{k}_o \cdot \bar{r})}e^{-i\omega t} \tag{7.74}$$

The first subscript shows the wave type (incident, reflected, or transmitted in Region 1). The reflected wave is

$$\bar{E}_{r,0} = \left[\hat{v}_0(k_{0z})R_{01VV}e^{i(\bar{k}_o \cdot \bar{r})}\right]e^{-i\omega t} \tag{7.75}$$

R_{01VV} is the reflection coefficient when incident and reflected waves are both vertically polarized. The transmitted wave in Region 1 can be expressed as

$$\bar{E}_{t,1} = \left[\hat{e}\left(k_{1z}^{ed}\right)T_{Ve}e^{i(\bar{\kappa}_1^e \cdot \bar{r})}\right]e^{-i\omega t} \tag{7.76}$$

T_{Ve} is the transmission coefficient when incident wave is vertically polarized and transmitted wave is an extraordinary wave. Repeating the same procedure to find reflection and transmission coefficients by applying boundary conditions at z = 0 gives

$$R_{01VV} = \frac{\varepsilon_1 k_{oz} - \varepsilon_0 k_{1z}^e}{\varepsilon_1 k_{oz} + \varepsilon_0 k_{1z}^e} \tag{7.77}$$

$$T_{Ve} = \frac{2\varepsilon_1 k_{oz}}{\left(\varepsilon_1 k_{oz} + \varepsilon_0 k_{1z}^e\right)} \tag{7.78}$$

7.2.1.3 Numerical Example

The Fresnel coefficients derived for the single layer uniaxially anisotropic medium for a vertically oriented optic axis are numerically calculated for Taconic TLY-5A

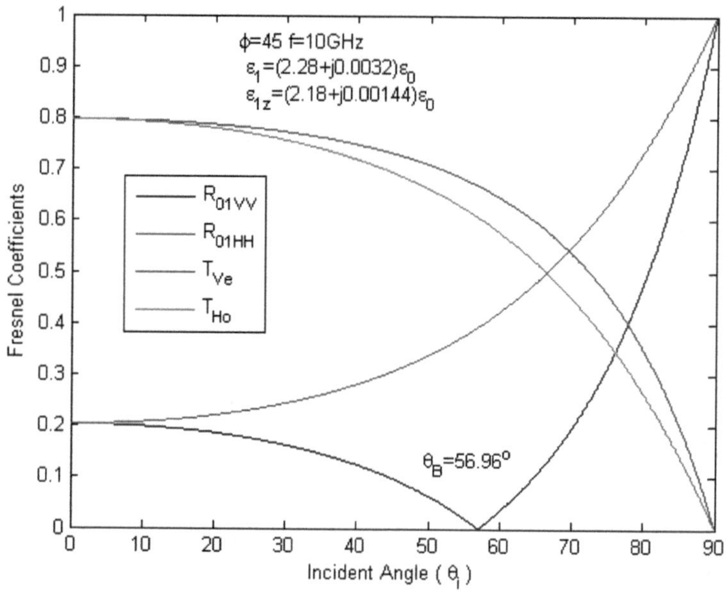

Fig. 7.4 Fresnel coefficients for single layer uniaxially anisotropic medium with a vertically oriented optic axis

material and illustrated in Fig. 7.4. TLY-5A is a negatively uniaxial anisotropic medium with dissipation factor of 0.0014 in (x, y)-direction and 0.00066 in z-direction at 10 GHz. The Brewster angle numerically calculated to be 56.96° as shown in Fig. 7.4.

The theoretical value of the Brewster angle is found from

$$\theta_B = a\sin\left(\sqrt{\frac{\varepsilon_{1z}(\varepsilon_1 - \varepsilon_0)}{\varepsilon_1\varepsilon_{1z} - \varepsilon_0^2}}\right)$$

The theoretical value is also found to be 56.96° as shown in Fig. 7.4.

7.2.2 Single-Layered Anisotropic Media: Optic Axis Tilted in One Direction

The cross polarization effect due to isotropic–anisotropic interface can be obtained if the optic axis of the uniaxially anisotropic medium is tilted. Cross polarization effect is very important in radiation and wave scattering problems. The geometry of the anisotropic medium is shown in Fig. 7.5 when its optic axis is tilted around

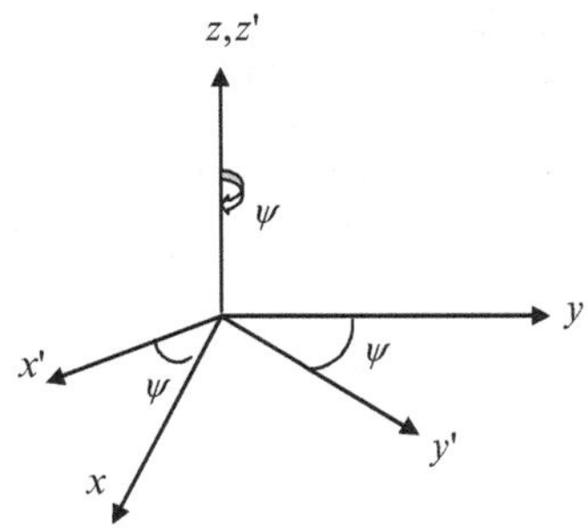

Fig. 7.5 Geormetry of a unaxially anisotropic medium with its optic axis tilted around z-axis by an angle ψ

z-direction by angle ψ. The permittivity tensor of the anisotropic medium with tilted optic axis becomes

$$\bar{\bar{\varepsilon}}^{oz} = \begin{bmatrix} \varepsilon_{11}^{oz} & 0 & 0 \\ 0 & \varepsilon_{22}^{oz} & \varepsilon_{23}^{oz} \\ 0 & \varepsilon_{32}^{oz} & \varepsilon_{33}^{oz} \end{bmatrix} \tag{7.79}$$

The first and second superscripts of the permittivity tensor indicate the direction where the optic axis is tilted. The matrix elements of the tensor in (7.79) are

$$\varepsilon_{11}^{oz} = \varepsilon_1 \tag{7.80}$$

$$\varepsilon_{22}^{oz} = \varepsilon_1 \cos^2\psi + \varepsilon_{1z}\sin^2\psi \tag{7.81}$$

$$\varepsilon_{23}^{oz} = \varepsilon_{32}^{oz} = (\varepsilon_1 - \varepsilon_{1z}) \cos\psi sin\psi \tag{7.82}$$

$$\varepsilon_{33}^{oz} = \varepsilon_1 \sin^2\psi + \varepsilon_{1z}\cos^2\psi \tag{7.83}$$

The single layer structure with the direction of wave vectors is illustrated in Fig. 7.6. We first analyze the wave propagation for TE wave incidence.

7.2.2.1 TE Wave

When the incident wave is TE wave, the electric field vector in Region 0 is expressed as

$$\bar{E}_{i,0} = \hat{h}_0(-k_{0z}) e^{i(\bar{\kappa}_o \cdot \bar{r})} e^{-i\omega t} \tag{7.84}$$

Figure 7.6 shows the wave vectors for the single layer structure when the incident wave is horizontally polarized. Upon reflection from the isotropic–anisotropic interface the reflected wave in Region 0 will have the cross polarized component R_{01HV}. Cross polarized components carries great importance in radiation and

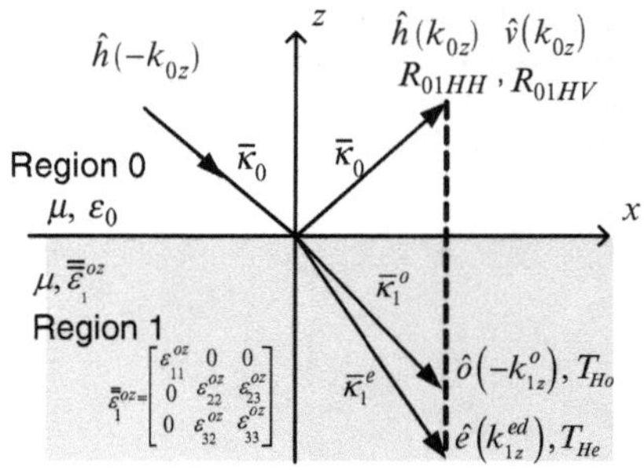

Fig. 7.6 Reflection and transmission of TE waves in single-layered anisotropic slab with a tilted optic axis in one direction

scattering problems. The first subscript of the cross polarized component represents the polarization of the incident wave whereas the second subscript represents the polarization of the reflected wave. The reflected electric field vector can be written as

$$\bar{E}_{r,0} = \left[\hat{h}_0(k_{0z})R_{01HH}e^{i(\bar{k}_o\cdot\bar{r})} + \hat{v}_0(k_{0z})R_{01HV}e^{i(\bar{k}_o\cdot\bar{r})}\right]e^{-i\omega t} \tag{7.85}$$

The transmitted wave also has the cross polarized component due to tilted optic axis. T_{He} is the transmitted cross polarized component and represents the transmitted extraordinary wave when the incident wave is TE wave. The transmitted field vector is

$$\bar{E}_{t,1} = \left[\hat{o}\left(-k_{1z}^o\right)T_{Ho}e^{i(\bar{\kappa}_1^o\cdot\bar{r})} + \hat{e}\left(k_{1z}^{ed}\right)T_{Ho}e^{i(\bar{\kappa}_1^e\cdot\bar{r})}\right]e^{-i\omega t} \tag{7.86}$$

Application of the boundary conditions given by (7.64) and (7.65) gives the refection and transmission coefficients as

$$R_{01HH} = -1 + \frac{T_{Ho}}{\alpha_d}\frac{1}{k_\rho}\left(k_\rho^2\cos\psi + k_y k_{1z}^o\sin\psi\right) - \frac{k_1^2}{k_\rho}k_x\sin\psi\frac{T_{He}}{\beta_d} \tag{7.87}$$

$$R_{01HV} = -\frac{T_{Ho}}{\alpha_d}\frac{k_o k_x}{k_\rho k_{oz}}k_{1z}^o\sin\psi + \frac{T_{He}}{\beta_d}\frac{k_o}{k_\rho k_{oz}}\left(k_\rho^2 k_{1z}^{ed}\cos\psi - k_y\left(k_{1z}^o\right)^2\sin\psi\right) \tag{7.88}$$

where

$$T_{He} = -\frac{\beta_d}{\lambda_e}\frac{2k_{oz}}{k_\rho}k_x\left(k_\rho^2 + k_{oz}k_{1z}^o\right)\sin\psi \tag{7.89}$$

$$T_{Ho} = -\frac{\alpha_d}{\lambda_e}\frac{2}{\left(k_{oz} + k_{1z}^o\right)}\frac{k_{oz}}{k_\rho} \times\left[k_\rho^2\left(k_1^2 k_{oz} - k_o^2 k_{1z}^{ed}\right)\cos\psi + k_y\left(k_o^2\left(k_{1z}^o\right)^2 - k_1^2 k_{oz}k_{1z}^{ed}\right)\sin\psi\right] \tag{7.90}$$

$$\lambda_e = \cos^2\psi k_\rho^2\left(k_1^2 k_{1z}^o - k_o^2 k_{1z}^{ed}\right) + \sin^2\psi\left[k_1^2\left(k_{oz} - k_{1z}^{ed}\right)\left(k_x^2 + k_{oz}k_{1z}^o\right) + k_y^2 k_{1z}^o\left(k_1^2 - k_o^2\right)\right]$$
$$+ \cos\psi\sin\psi k_y\left(k_{1z}^o + k_{oz}\right)\left(k_{1z}^o - k_{1z}^{ed}\right)\left(k_\rho^2 + k_{1z}^o k_{oz}\right) \tag{7.91}$$

$$\beta_d = \left[\frac{\varepsilon_1}{\varepsilon_1 - \varepsilon_{1z}}\left(k_x^2 + k_y^2 + \left(k_{1z}^{ed}\right)^2 - \omega^2\mu\varepsilon_1\right)\left(k_x^2 + k_y^2 + \left(k_{1z}^{ed}\right)^2 - \omega^2\mu(\varepsilon_1 + \varepsilon_{1z})\right)\right]^{1/2} \tag{7.92}$$

$$\alpha_d = \left[k_x^2 + \left(k_y\cos\psi + k_{1z}^o\sin\psi\right)^2\right]^{1/2} \tag{7.93}$$

7.2.2.2 TM Wave

Figure 7.7 illustrates the single layer structure when the incident wave is vertically polarized. The incident electric field vector can be written as

$$\bar{E}_{i,0} = \hat{v}_0(-k_{0z})e^{i(\bar{\kappa}_o\cdot\bar{r})}e^{-i\omega t} \tag{7.94}$$

The reflected wave with the cross polarized term is R_{01VH} is

$$\bar{E}_{r,0} = \left[\hat{v}_0(k_{0z})R_{01VV}e^{i(\bar{k}_o\cdot\bar{r})} + \hat{h}_0(k_{0z})R_{01VH}e^{i(\bar{k}_o\cdot\bar{r})}\right]e^{-i\omega t} \tag{7.95}$$

R_{01VH} indicates that the incident wave is vertically polarized or TM wave, and the reflected wave is horizontally polarized or TE wave. The transmitted wave in Region 1 can be expressed as

$$\bar{E}_{t,1} = \left[\hat{o}\left(-k_{1z}^o\right)T_{Vo}e^{i(\bar{\kappa}_1^o\cdot\bar{r})} + \hat{e}\left(k_{1z}^{ed}\right)T_{Ve}e^{i(\bar{\kappa}_1^e\cdot\bar{r})}\right]e^{-i\omega t} \tag{7.96}$$

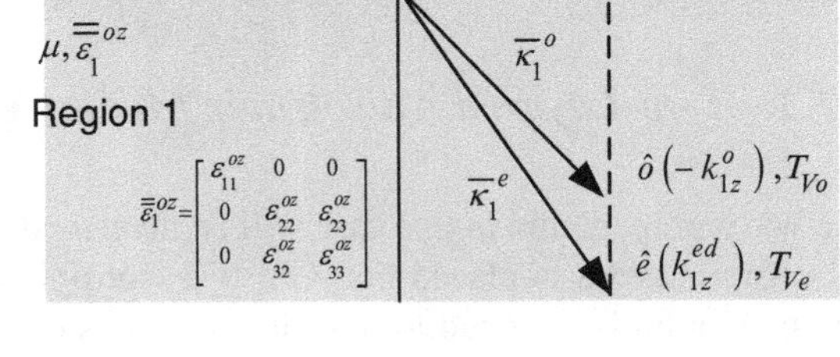

Fig. 7.7 Reflection and transmission of TM waves in single-layered anisotropic slab with a tilted optic axis in one direction

T_{Vo} and T_{Ve} are the transmission coefficients when the incident wave is TM wave and transmitted wave ordinary wave or extraordinary wave, respectively. The reflection and transmission coefficients are found using boundary conditions as described before and given as

$$R_{01VV} = 1 - \frac{T_{Vo}}{\alpha_d}\frac{k_o k_x}{k_\rho k_{oz}} k_{1z}^o \sin\psi + \frac{T_{Ve}}{\beta_d}\frac{k_o}{k_\rho k_{oz}}\left(k_\rho^2 k_{1z}^{ed}\cos\psi - k_y\left(k_{1z}^o\right)^2\sin\psi\right) \quad (7.97)$$

$$R_{01VH} = \frac{T_{Vo}}{\alpha_d}\frac{1}{k_\rho}\left(k_\rho^2\cos\psi + k_y k_{1z}^o \sin\psi\right) - \frac{T_{Ve}}{\beta_d}\frac{k_1^2}{k_\rho}k_x\sin\psi \quad (7.98)$$

where

$$T_{Ve} = \frac{\beta_d}{\lambda_e}\frac{2k_o}{k_\rho}\left(k_\rho^2 k_{oz}\cos\psi + k_y k_{oz} k_{1z}^o \sin\psi\right) \quad (7.99)$$

$$T_{Vo} = \frac{\alpha_d}{\lambda_e}\frac{2}{\left(k_{oz} + k_{1z}^o\right)}\frac{k_{oz}}{k_\rho}\left[k_1^2 k_x k_o\left(k_{oz} - k_{1z}^{ed}\right)\sin\psi\right] \quad (7.100)$$

The coefficients α_d, β_d, and λ_d are given by (7.91)–(7.93). The transmission and reflection coefficients calculated for TE and TM waves given by (7.87)–(7.90) and (7.97)–(7.100) check with the results given in [3].

7.2.2.3 Numerical Example

The Fresnel coefficients derived for the single layer vertically uniaxially anisotropic medium with a tilted optic axis are numerically calculated for Taconic TLY-5A material and illustrated in Fig. 7.8.

As illustrated, the cross polarization components exist because of the isotropic–anisotropic interface. The magnitude of the cross polarized components varies with the tilt angle and observation angle. The cross polarized terms are very important due to their contribution on the electric and magnetic field intensities in radiation and scattering problems.

7.2.3 Two-Layered Anisotropic Media: Vertically Uniaxial Case

The wave propagation in this case will be investigated when two-layered uniaxially anisotropic media is placed between two isotropic regions. The uniaxially anisotropic medium is assumed to have its optic axis oriented in z-direction.

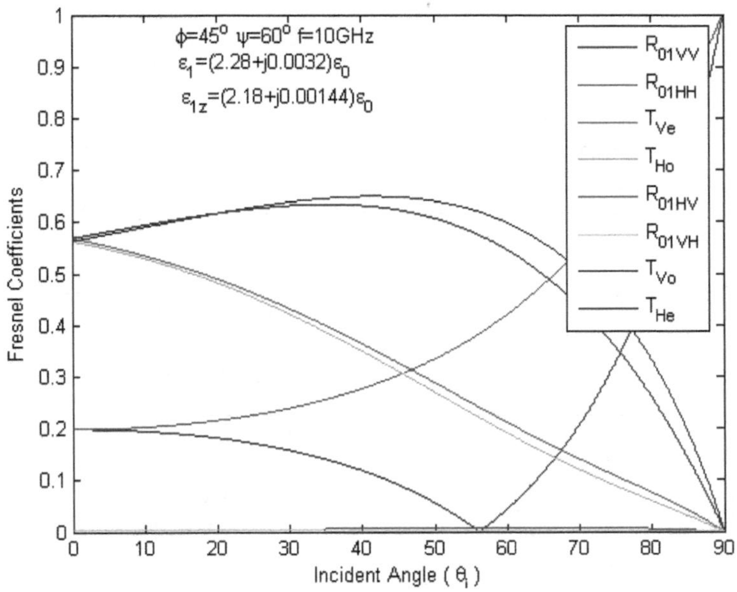

Fig. 7.8 Fresnel coefficients for single layer uniaxially anisotropic medium with a tilted optic axis

7.2.3.1 TE Wave

The two-layered anisotropic structure when the incident wave is horizontally polarized is shown in Fig. 7.9. The electric field vector in Region 0 is

$$\bar{E}_0 = \left[\hat{h}_0(-k_{0z})e^{i(\bar{\kappa}_o\cdot\bar{r})} + \hat{h}_0(k_{0z})R_H e^{i(\bar{k}_o\cdot\bar{r})}\right]e^{-i\omega t} \tag{7.101}$$

The reflected wave is also a horizontally polarized wave. R_H is the reflection coefficient which has the reflected wave components from isotropic–anisotropic interface and anisotropic-isotropic interface when incident wave is horizontally polarized. The transmitted wave in Region 1 is expressed as

$$\bar{E}_1 = \left[\hat{o}\left(-k_{1z}^o\right)T_{Ho1}e^{i(\bar{\kappa}_1^o\cdot\bar{r})} + \hat{o}\left(k_{1z}^o\right)T_{Ho2}e^{i(\bar{k}_1^o\cdot\bar{r})}\right]e^{-i\omega t} \tag{7.102}$$

T_{Ho1} is the transmission coefficient for the downward propagating ordinary wave and T_{Ho2} is transmission coefficient for upward propagating ordinary wave. In Region 2, there is only downward propagating wave which is horizontally polarized. The field vector in Region 2 can be expressed as

$$\bar{E}_2 = \left[\hat{h}_2\left(-k_{2z}^o\right)T_{H2}e^{i(\bar{\kappa}_2^o\cdot\bar{r})}\right]e^{-i\omega t} \tag{7.103}$$

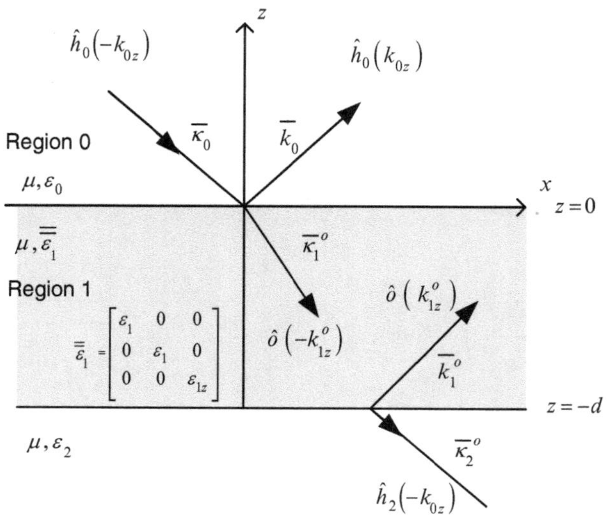

Fig. 7.9 Reflection and transmission of TE waves in two-layered anisotropic structure

T_{H2} is the transmission coefficient in Region 2 when the incident wave in Region 0 is a horizontally polarized wave . The Fresnel coefficients R_H, T_{Ho1}, T_{Ho2} and T_{H2} are found by using the following boundary conditions at $z = 0, -d$

$$\hat{z} \times \bar{E}_0 = \hat{z} \times \bar{E}_1 \tag{7.104}$$

$$\hat{z} \times \nabla \times \bar{E}_0 = \hat{z} \times \nabla \times \bar{E}_1 \tag{7.105}$$

$$\hat{z} \times \bar{E}_1 = \hat{z} \times \bar{E}_2 \tag{7.106}$$

$$\hat{z} \times \nabla \times \bar{E}_1 = \hat{z} \times \nabla \times \bar{E}_2 \tag{7.107}$$

The Fresnel coefficients after application of the boundary conditions (7.104) and (7.105) are found as

$$R_H = \frac{\left(R_{H01} + R_{H12} e^{i2k_{1z}d}\right)}{\alpha_{H2}} \tag{7.108}$$

$$T_{Ho1} = \frac{(1 + R_{H01})}{\alpha_{H2}} \tag{7.109}$$

$$T_{Ho2} = \frac{(1 + R_{H01}) R_{H12} e^{i2k_{1z}d}}{\alpha_{H2}} \tag{7.110}$$

$$T_{H2} = \frac{(1 + R_{H01})(1 + R_{H12}) e^{i(k_{1z} - k_{2z})d}}{\alpha_{H2}} \tag{7.111}$$

where

$$R_{H01} = \frac{k_{0z} - k_{1z}}{k_{0z} + k_{1z}} \tag{7.112}$$

$$R_{H12} = \frac{k_{1z} - k_{2z}}{k_{1z} + k_{2z}} \tag{7.113}$$

$$\alpha_{H2} = 1 + R_{H01} R_{H12} e^{i2k_{1z}d} \tag{7.114}$$

The new unit vectors, wave vector and the dispersion relation in Region 2 are

$$\hat{h}_2(\pm k_{2z}) = \frac{\hat{z} \times \bar{k}_2}{k_\rho} = \frac{(-\hat{x}k_y + \hat{y}k_x)}{k_\rho} \tag{7.115}$$

$$\hat{o}(k^o_{1z}) = \frac{\hat{z} \times \bar{k}^o_1}{|\hat{z} \times \bar{k}^o_1|} = \frac{(-\hat{x}k_y + \hat{y}k_x)}{k_\rho} \tag{7.116}$$

$$\bar{\kappa}^o_2 = \bar{k}_\rho - k^\circ_{2z}\hat{z}$$

$$k_2 = \omega\sqrt{\mu\varepsilon_2} \tag{7.117}$$

7.2.3.2 TM Wave

Now, assume the incident wave in Region 0 is TM wave for the layered structure shown in Fig. 7.10. The field vectors in Region 0, 1, and 2 can be expressed as

$$\bar{E}_0 = \left[\hat{v}_0(-k_{0z})e^{i(\bar{\kappa}_o\cdot\bar{r})} + \hat{v}_0(k_{0z})R_V e^{i(\bar{k}_o\cdot\bar{r})}\right]e^{-i\omega t} \tag{7.118}$$

$$\bar{E}_1 = \left[\hat{e}\left(k^{ed}_{1z}\right)T_{Ve1}e^{i(\bar{\kappa}^e_1\cdot\bar{r})} + \hat{e}\left(k^{eu}_{1z}\right)T_{Ve2}e^{i(\bar{k}^e_1\cdot\bar{r})}\right]e^{-i\omega t} \tag{7.119}$$

$$\bar{E}_2 = \left[\hat{v}_2\left(-k^o_{2z}\right)T_{V2}e^{i(\bar{\kappa}^o_2\cdot\bar{r})}\right]e^{-i\omega t} \tag{7.120}$$

The reflection and transmission coefficients R_V, T_{Ve1}, T_{Ve2} and T_{V2} are found using the boundary conditions given by (7.104)–(7.107) by for the electric field vectors given by (7.118)–(7.120). The Fresnel coefficients are found with application of boundary conditions similar to TE wave and given as

$$R_V = \frac{\left(R_{V01} + R_{V12}e^{i2k_{1z}d}\right)}{\alpha_{V2}} \tag{7.121}$$

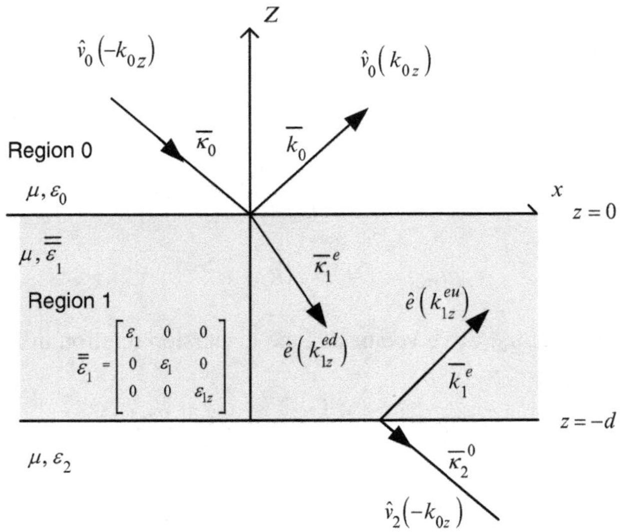

Fig. 7.10 Reflection and transmission of TM waves in two-layered anisotropic structure

$$T_{Ve1} = \frac{(1 + R_{V01})}{\alpha_{V2}} \frac{k_0 k_{1u}}{k_1^2} \frac{\varepsilon_1}{\varepsilon_{1z}} \tag{7.122}$$

$$T_{Ve2} = \frac{(1 + R_{V01})}{\alpha_{V2}} \frac{k_0 k_{1u}}{k_1^2} \frac{\varepsilon_1}{\varepsilon_{1z}} (1 + R_{V12}) e^{i2k_{1z}^e d} \tag{7.123}$$

$$T_{V2} = \frac{k_0}{k_2} \frac{(1 + R_{V01})}{\alpha_{V2}} (1 + R_{V12}) e^{i\left(k_{1z}^e - k_{2z}\right)d} \tag{7.124}$$

where

$$R_{V01} = \frac{\varepsilon_1 k_{0z} - \varepsilon_0 k_{1z}^e}{\varepsilon_1 k_{0z} + \varepsilon_0 k_{1z}^e} \tag{7.125}$$

$$R_{V12} = \frac{\varepsilon_2 k_{1z}^e - \varepsilon_1 k_{2z}}{\varepsilon_2 k_{1z}^e + \varepsilon_1 k_{2z}} \tag{7.126}$$

$$\alpha_{V2} = 1 + R_{V01} R_{V12} e^{i2k_{1z}^e d} \tag{7.127}$$

7.2.3.3 Application Example: Microstrip with Anisotropic Medium

Application of the electromagnetic wave propagation problem through composite structure can be best illustrated with making the second region perfect conductor.

When the permittivity of the second layer taken to be infinitely large, $\varepsilon_2 \rightarrow \infty$, it can be approximated with a perfect conductor.

The final configuration with this approximation can be considered as microstrip structure as shown in Fig. 7.11. The dielectric layer in this microstrip structure is a vertically uniaxial anisotropic medium. This configuration can be used for radiation or scattering problems. The calculations for radiation and scattering require finding Dyadic Green's Functions (DGF) of the structure. DGF of the medium has the same form as the electric field in that region as described in Chap. 5. As a result, Fresnel coefficients in electric field vectors have to be used for the region under consideration to be able to construct DGF. For the microstrip involving uniaxially anisotropic medium shown in Fig. 7.11, the Fresnel coefficients for TE and TM waves derived in Sect. 7.2.3 are modified with the substitution of $\varepsilon_2 \rightarrow \infty$ into (7.108)–(7.114) and (7.121)–(7.127). The modified Fresnel coefficients for TE waves are

$$R_H \rightarrow \frac{\left(R_{H01} - e^{i2k_{1z}d}\right)}{\alpha_{H2}} \tag{7.128}$$

$$T_{Ho1} = \frac{(1 + R_{H01})}{\alpha_{H2}} \tag{7.129}$$

$$T_{Ho2} \rightarrow -\frac{(1 + R_{H01})e^{i2k_{1z}d}}{\alpha_{H2}} \tag{7.130}$$

$$T_{H2} \rightarrow 0 \tag{7.131}$$

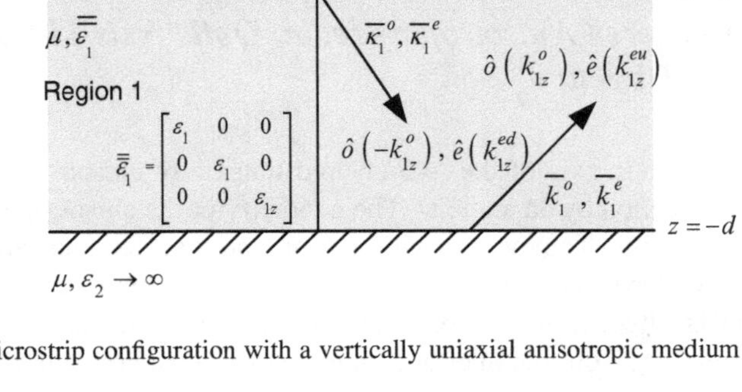

Fig. 7.11 Microstrip configuration with a vertically uniaxial anisotropic medium

where

$$R_{H01} = \frac{k_{0z} - k_{1z}}{k_{0z} + k_{1z}} \tag{7.132}$$

$$R_{H12} \to -1 \tag{7.133}$$

$$\alpha_{H2} \to 1 - R_{H01} e^{i2k_{1z}d} \tag{7.134}$$

Similarly, the modified Fresnel coefficients for TE waves are found as

$$R_V \to \frac{\left(R_{V01} + e^{i2k_{1z}d}\right)}{\alpha_{V2}} \tag{7.135}$$

$$T_{Ve1} = \frac{\left(1 + R_{V01}\right)}{\alpha_{V2}} \frac{k_0 k_{1u}}{k_1^2} \frac{\varepsilon_1}{\varepsilon_{1z}} \tag{7.136}$$

$$T_{Ve2} \to \frac{\left(1 + R_{V01}\right)}{\alpha_{V2}} \frac{k_0 k_{1u}}{k_1^2} \frac{\varepsilon_1}{\varepsilon_{1z}} 2e^{i2k_{1z}^e d} \tag{7.137}$$

$$T_{V2} \to 0 \tag{7.138}$$

where

$$R_{V01} = \frac{\varepsilon_1 k_{0z} - \varepsilon_0 k_{1z}^e}{\varepsilon_1 k_{0z} + \varepsilon_0 k_{1z}^e} \tag{7.139}$$

$$R_{V12} \to 1 \tag{7.140}$$

$$\alpha_{V2} \to 1 + R_{V01} e^{i2k_{1z}^e d} \tag{7.141}$$

7.2.4 Two-Layered Anisotropic Media: Optic Axis Tilted in One Direction

Let's assume the optic axis of the two-layered uniaxially anisotropic medium is tilted around z-direction by an angle ψ. The geometry of the anisotropic medium is illustrated in Fig. 7.5. The permittivity tensor and its elements are given by (7.79)–(7.81). We have seen in Sect. 7.2.2 that when the optic axis of the anisotropic medium is tilted, it allows existence of the cross polarization components across the isotropic–anisotropic interface. In the two-layered configuration, we have anisotropic-isotropic interface in addition to the isotropic–anisotropic interface across the boundaries at z = 0, and z = −d. Although the calculation of the Fresnel coefficients gets complicated and tedious, its straight forward with the method proposed in this chapter by application of the boundary conditions as described.

7.2.4.1 TE Wave

When the incident wave is TE wave or horizontally polarized wave, $\hat{h}_o(-k_{oz})$, the electric field vector in Region 0 will have both horizontally polarized and vertically polarized components due to isotropic–anisotropic interface as discussed in Sect. 7.2.2. The field vector in Region 0 can be expressed as

$$\bar{E}_0 = \left[\hat{h}_0(-k_{0z})e^{i(\bar{\kappa}_0 \cdot \bar{r})} + \hat{h}_0(k_{0z})R_{01HH}e^{i(\bar{k}_o \cdot \bar{r})} + \hat{v}_0(k_{0z})R_{01HV}e^{i(\bar{k}_o \cdot \bar{r})} \right] e^{-i\omega t} \quad (7.142)$$

Upon transmission from Region 0 to Region 1, the transmitted horizontally polarized wave is decomposed into downward propagating ordinary and extraordinary wave components, $\hat{o}(-k_{1z}^o)$ and $\hat{e}(k_{1z}^{ed})$, with transmission coefficients T_{Ho} and T_{He}. The ordinary and extraordinary downward propagating wave components will be reflected back from the anisotropic–isotropic interface as illustrated in Fig. 7.12. The wave components reflected from anisotropic-isotropic interface will be upward propagating and are designated by unit vectors $\hat{o}(k_{1z}^o)$ and $\hat{e}(k_{1z}^{eu})$

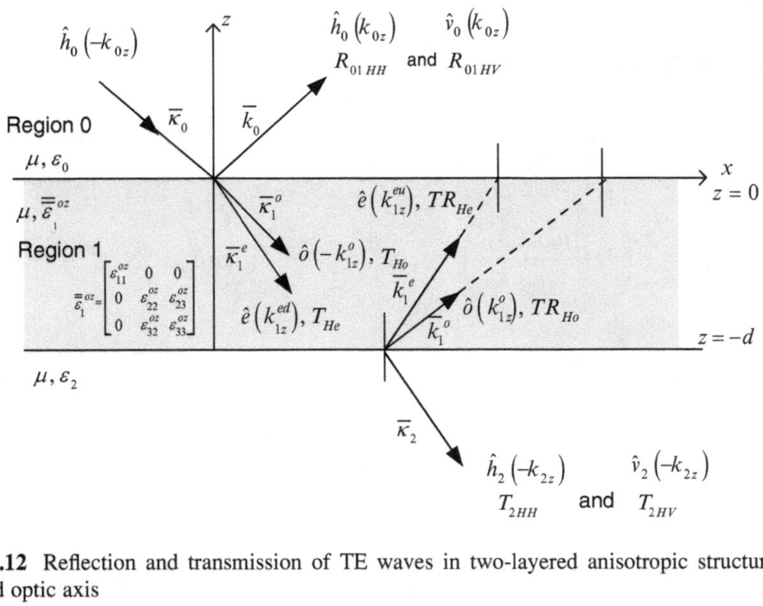

Fig. 7.12 Reflection and transmission of TE waves in two-layered anisotropic structure with a tilted optic axis

with the transmission-reflection coefficients TR_{Ho} and TR_{He}. The electric field vector in Region 1 can be written as

$$\bar{E}_1 = \begin{bmatrix} T_{Ho}\hat{o}(-k_{1z}^{o})e^{i(\bar{\kappa}_1^{o}\cdot\bar{r})} + TR_{Ho}\hat{o}(k_{1z}^{o})e^{i(\bar{k}_1^{o}\cdot\bar{r})} \\ +T_{He}\hat{e}(k_{1z}^{ed})e^{i(\bar{\kappa}_1^{e}\cdot\bar{r})} + TR_{He}\hat{e}(k_{1z}^{eu})e^{i(\bar{k}_1^{e}\cdot\bar{r})} \end{bmatrix} e^{-i\omega t} \tag{7.143}$$

The transmitted wave in Region 2 has both horizontally and vertically polarized waves due to anisotropic-isotropic interface. The transmitted wave is represented with transmission coefficients T_{2HH} and T_{2HV}. The field vector in Region 2 is

$$\bar{E}_2 = \left[\hat{h}_2(-k_{2z})T_{2HH}e^{i(\bar{\kappa}_2\cdot\bar{r})} + \hat{v}_2(-k_{2z})T_{2HH}e^{i(\bar{\kappa}_2\cdot\bar{r})}\right] e^{-i\omega t} \tag{7.144}$$

The Fresnel coefficients are found by application boundary conditions given using (7.104)–(7.107) as described in the previous sections. TM wave analysis can be carried out by considering the incident wave as vertically polarized wave.

7.2.5 *Multilayered Anisotropic Media*

Consider the following stratified general anisotropic medium shown in Fig. 7.13. Assume that the principal axes of the medium are oriented parallel to the coordinate axes. This configuration is practical and can be adjusted to have composite planar multilayer circuit by changing the permittivity and/or permeability parameters of the medium in each layer.

7.2.5.1 TE Wave

From continuity of the tangential components of the field vectors along the boundary at $x = -d_l$, we relate the wave components as

$$A_l e^{-ik_{lx}d_l} = \frac{1}{2}\left(1 + \frac{\mu_{lz}k_{(l+1)x}}{\mu_{(l+1)}k_{lx}}\right)\left(A_{l+1}e^{-ik_{(l+1)x}d_l} + R_{l(l+1)}B_{l+1}e^{ik_{(l+1)x}d_l}\right) \tag{7.145}$$

$$B_l e^{-ik_{lx}d_l} = \frac{1}{2}\left(1 + \frac{\mu_{lz}k_{(l+1)x}}{\mu_{(l+1)}k_{lx}}\right)\left(R_{l(l+1)}A_{l+1}e^{-ik_{(l+1)x}d_l} + B_{l+1}e^{ik_{(l+1)x}d_l}\right) \tag{7.146}$$

The wave components can now be related through wave components in the $(l+1)$ as

$$\begin{pmatrix} A_{l+1}e^{-ik_{(l+1)x}d_{l+1}} \\ B_{l+1}e^{ik_{(l+1)x}d_{l+1}} \end{pmatrix} = F_{(l+1)l}\begin{pmatrix} A_l e^{-ik_{lx}d_l} \\ B_l e^{ik_{lx}d_l} \end{pmatrix} \tag{7.147}$$

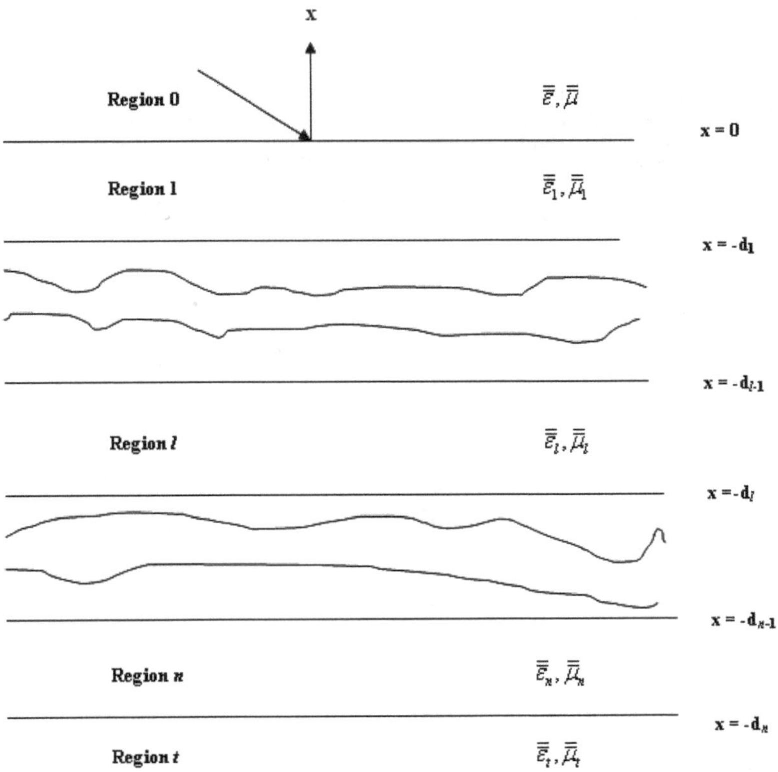

Fig. 7.13 Geometry of multilayered anisotropic media

where the forward propagation matrix is

$$F_{(l+1)l} = \frac{1}{2}\left(1+\Gamma^{e}_{(l+1)l}\right)\begin{pmatrix} e^{-ik_{(l+1)x}(d_{l+1}-d_l)} & R_{(l+1)l}e^{-ik_{(l+1)x}(d_{l+1}-d_l)} \\ R_{(l+1)l}e^{ik_{(l+1)x}(d_{l+1}-d_l)} & e^{ik_{(l+1)x}(d_{l+1}-d_l)} \end{pmatrix} \tag{7.148}$$

$$R_{(l+1)l} = -R_{l(l+1)} = \frac{\left(1-\Gamma^{e}_{(l+1)l}\right)}{\left(1+\Gamma^{e}_{(l+1)l}\right)} \tag{7.149}$$

and

$$\Gamma^{e}_{(l+1)l} = \frac{\mu_{(l+1)z}k_{lx}}{\mu_{lz}k_{(l+1)x}} \tag{7.150}$$

The analysis and the structure of the forward propagation matrix is similar to the one that is found for the multilayer isotropic structure. Once the forward propagation matrix is found, the reflection and transmission coefficients can be easily calculated. TM wave analysis is done by applying the duality relations.

References

1. J.A. Kong, Electromagnetic Wave Theory, EMW Publishing, Cambridge, MA, 2000.
2. L. Tsang, E. Njoku, and J.A. Kong, "Microwave thermal emission from stratified medium with nonuniform temperature distribution," J. Appl. Phys., vol. 46, no. 12, pp. 5127–5133, Dec. 1975.
3. J.K. Lee, and J.A. Kong, "Dyadic Green's functions for layered anisotropic medium," Electromagnetics, vol. 3, pp.111–130, Apr–June 1983.

Chapter 8
Microwave Devices Using Anisotropic and Gyrotropic Media

In this chapter, the theory that is introduced in the previous chapters for isotropic, anisotropic and gyrotropic materials finds an application. Application examples using isotropic, anisotropic and gyrotropic materials will be introduced with detail. Most of the design examples have analytical stage involving theory of the operation, simulation stage for verification of the design and experimental stage for implementation. The device performance using these three types of materials will be compared to understand the effect of anisotropy or nonreciprocity. The devices that will be analyzed, simulated and designed including waveguides, directional couplers, microwave filters, and nonreciprocal phase shifters.

8.1 Waveguide Design

In this section, rectangular waveguide design procedure for isotropic, gyrotropic and anisotropic media will be given. The wave propagation in waveguide can be analyzed by using two curl equations given in Chap. 1 in the absence of electric and magnetic current density as

$$\nabla \times \bar{E} = i\omega\bar{\bar{\mu}} \cdot \bar{H} \tag{8.1}$$

$$\nabla \times \bar{H} = -i\omega\bar{\bar{\varepsilon}} \cdot \bar{E} \tag{8.2}$$

where $\nabla = \left(\hat{x}\frac{\partial}{\partial x} + \hat{y}\frac{\partial}{\partial y} + \hat{z}\frac{\partial}{\partial z}\right)$. Expansion of (8.1) and (8.2) for electric and magnetic field components give two sets of equations.

$$\frac{\partial E_z}{\partial y} - \frac{\partial E_y}{\partial z} = i\omega\left[\mu_{11}H_x + \mu_{12}H_y + \mu_{13}H_z\right] \tag{8.3}$$

$$\frac{\partial E_x}{\partial z} - \frac{\partial E_z}{\partial x} = i\omega\left[\mu_{21}H_x + \mu_{22}H_y + \mu_{23}H_z\right] \tag{8.4}$$

A. Eroglu, *Wave Propagation and Radiation in Gyrotropic and Anisotropic Media*, DOI 10.1007/978-1-4419-6024-5_8, © Springer Science+Business Media, LLC 2010

$$\frac{\partial E_y}{\partial x} - \frac{\partial E_x}{\partial y} = i\omega\left[\mu_{31}H_x + \mu_{32}H_y + \mu_{33}H_z\right] \tag{8.5}$$

and

$$\frac{\partial H_z}{\partial y} - \frac{\partial H_y}{\partial z} = -i\omega\left[\varepsilon_{11}E_x + \varepsilon_{12}E_y + \varepsilon_{13}E_z\right] \tag{8.6}$$

$$\frac{\partial H_x}{\partial z} - \frac{\partial H_z}{\partial x} = -i\omega\left[\varepsilon_{21}E_x + \varepsilon_{22}E_y + \varepsilon_{23}E_z\right] \tag{8.7}$$

$$\frac{\partial H_y}{\partial x} - \frac{\partial H_x}{\partial y} = -i\omega\left[\varepsilon_{31}E_x + \varepsilon_{32}E_y + \varepsilon_{33}E_z\right] \tag{8.8}$$

The geometry of the rectangular waveguide is given in Fig. 8.1. The guide walls are assumed to be perfect conductor with $\{\sigma_c\} = \infty$. It is assumed that the wave propagation is in the z-direction and the electric and magnetic field vector components are functions of x and y. Then, the electric and magnetic field vector amplitudes can be written in the following form

$$\bar{F} = \bar{F}(x, y)e^{ikz} \tag{8.9}$$

where

$$\bar{F}(x, y) = F_x(x, y)\hat{x} + F_y(x, y)\hat{y} + F_z(x, y)\hat{z} \tag{8.10}$$

F represent electric field intensity E or magnetic field intensity H. The time dependence $e^{-i\omega t}$ is assumed as previously done. The analysis will be based on the lossless dielectric filling.

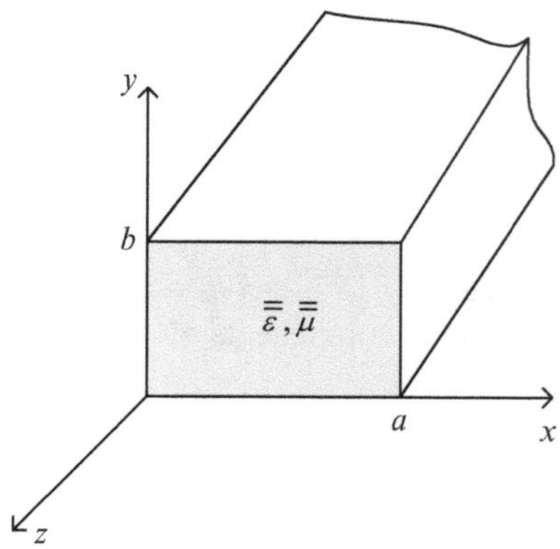

Fig. 8.1 Geometry of the rectangular waveguide

8.1.1 Waveguide Design with Isotropic Media

Consider the rectangular waveguide illustrated in Fig. 8.1. Assume that the dielectric filling inside the guide is isotropic with permittivity constant ε and permeability constant μ. So that

$$\bar{\bar{\varepsilon}} = \varepsilon\bar{\bar{I}} \quad \text{and} \quad \bar{\bar{\mu}} = \mu\bar{\bar{I}} \tag{8.11}$$

When (8.11) is substituted into (8.3)–(8.8) with (8.9), they take the following form

$$-ikE_y + \frac{\partial E_z}{\partial y} = i\omega\mu H_x \tag{8.12}$$

$$ikE_x - \frac{\partial E_z}{\partial x} = i\omega\mu H_y \tag{8.13}$$

$$\frac{\partial E_y}{\partial x} - \frac{\partial E_x}{\partial y} = i\omega\mu H_z \tag{8.14}$$

and

$$-ikH_y + \frac{\partial H_z}{\partial y} = -i\omega\varepsilon E_x \tag{8.15}$$

$$ikH_x - \frac{\partial H_z}{\partial x} = -i\omega\varepsilon E_y \tag{8.16}$$

$$\frac{\partial H_y}{\partial x} - \frac{\partial H_x}{\partial y} = -i\omega\varepsilon E_z \tag{8.17}$$

Now, we express transverse components of the field vectors, E_x, E_y, and H_x, H_y, in terms of the longitudinal components, E_z, H_z, which are the field vectors in the direction of propagation. By eliminating H_x from (8.12) and (8.16), and H_y from (8.13) and (8.15), we obtain

$$\mathrm{E}_y = -\frac{1}{(\omega^2\mu\varepsilon - k^2)}\left[ik\frac{\partial E_z}{\partial y} - i\omega\mu\frac{\partial H_z}{\partial x}\right] \tag{8.18}$$

$$\mathrm{E}_x = -\frac{1}{(\omega^2\mu\varepsilon - k^2)}\left[ik\frac{\partial E_z}{\partial x} + i\omega\mu\frac{\partial H_z}{\partial y}\right] \tag{8.19}$$

and

$$H_y = -\frac{1}{(\omega^2\mu\varepsilon - k^2)}\left[ik\frac{\partial H_z}{\partial y} + i\omega\varepsilon\frac{\partial E_z}{\partial x}\right] \tag{8.20}$$

$$H_x = -\frac{1}{(\omega^2\mu\varepsilon - k^2)}\left[ik\frac{\partial H_z}{\partial x} - i\omega\varepsilon\frac{\partial E_z}{\partial y}\right] \tag{8.21}$$

8.1.1.1 TE$_{\mathrm{mn}}$ Modes

For TE modes, the longitudinal component of the electric filed should be zero, $E_z = 0$. Application of this constraint on the transverse components of the electric and magnetic field vectors given by (8.18)–(8.21) reduces them to

$$\mathrm{E}_y = \frac{i\omega\mu}{(\omega^2\mu\varepsilon - k^2)}\frac{\partial H_z}{\partial x} \tag{8.22}$$

$$\mathrm{E}_x = -\frac{i\omega\mu}{(\omega^2\mu\varepsilon - k^2)}\frac{\partial H_z}{\partial y} \tag{8.23}$$

and

$$H_y = -\frac{ik}{(\omega^2\mu\varepsilon - k^2)}\frac{\partial H_z}{\partial y} \tag{8.24}$$

$$H_x = -\frac{ik}{(\omega^2\mu\varepsilon - k^2)}\frac{\partial H_z}{\partial x} \tag{8.25}$$

When we take the derivative of (8.23), and (8.24) with respect to y and x, respectively

$$\frac{\partial E_y}{\partial x} = \frac{i\omega\mu}{(\omega^2\mu\varepsilon - k^2)}\frac{\partial^2 H_z}{\partial x^2} \tag{8.26}$$

$$\frac{\partial E_x}{\partial y} = -\frac{i\omega\mu}{(\omega^2\mu\varepsilon - k^2)}\frac{\partial^2 H_z}{\partial y^2} \tag{8.27}$$

and subtract (8.27) from (8.26), we obtain

$$\frac{\partial E_y}{\partial x} - \frac{\partial E_x}{\partial y} = \frac{i\omega\mu}{(\omega^2\mu\varepsilon - k^2)}\frac{\partial^2 H_z}{\partial x^2} + \frac{i\omega\mu}{(\omega^2\mu\varepsilon - k^2)}\frac{\partial^2 H_z}{\partial y^2}$$

or

$$\frac{\partial E_y}{\partial x} - \frac{\partial E_x}{\partial y} = \frac{i\omega\mu}{(\omega^2\mu\varepsilon - k^2)}\left[\frac{\partial^2 H_z}{\partial x^2} + \frac{\partial^2 H_z}{\partial y^2}\right] \tag{8.28}$$

Since from (8.14),

$$\frac{\partial E_y}{\partial x} - \frac{\partial E_x}{\partial y} = i\omega\mu H_z \tag{8.29}$$

then,

$$\frac{i\omega\mu}{(\omega^2\mu\varepsilon - k^2)}\left[\frac{\partial^2 H_z}{\partial x^2} + \frac{\partial^2 H_z}{\partial y^2}\right] = i\omega\mu H_z$$

or

$$\frac{\partial^2 H_z}{\partial x^2} + \frac{\partial^2 H_z}{\partial y^2} + (\omega^2\mu\varepsilon - k^2)H_z = 0 \tag{8.30}$$

Same procedure can be applied to find wave equation for TM waves. The solution of the second order differential equation given in (8.30) takes the following form [1].

$$H_z(x, y) = A\cos(k_x x)\cos(k_y y) \tag{8.31}$$

Application of boundary conditions at $x = 0, a$ and $y = 0, b$ requires tangential components of the magnetic field vectors to be zero, i.e. $H_z = 0$. When the boundary conditions are applied, we find from (8.31) that

$$k_x = \frac{m\pi}{a}, \quad m = 0, 1, 2... \tag{8.32}$$

$$k_y = \frac{n\pi}{b}, \quad n = 0, 1, 2... \tag{8.33}$$

So, (8.31) can be written as

$$H_z(x, y) = A\cos\left(\frac{m\pi}{a}x\right)\cos\left(\frac{n\pi}{b}y\right) \tag{8.34}$$

The other components of the electric field vectors are found from (8.22)–(8.25)

$$E_x(x, y) = A\frac{-i\omega\mu}{(\omega^2\mu\varepsilon - k^2)}\frac{n\pi}{b}\cos\left(\frac{m\pi}{a}x\right)\sin\left(\frac{n\pi}{b}y\right) \tag{8.35}$$

$$E_y(x, y) = A\frac{i\omega\mu}{(\omega^2\mu\varepsilon - k^2)}\frac{m\pi}{a}\sin\left(\frac{m\pi}{a}x\right)\cos\left(\frac{n\pi}{b}y\right) \tag{8.36}$$

$$H_x(x,y) = A\frac{-ik}{(\omega^2\mu\varepsilon - k^2)}\frac{m\pi}{a}\sin\left(\frac{m\pi}{a}x\right)\cos\left(\frac{n\pi}{b}y\right) \tag{8.37}$$

$$H_y(x,y) = A\frac{-ik}{(\omega^2\mu\varepsilon - k^2)}\frac{n\pi}{b}\cos\left(\frac{m\pi}{a}x\right)\sin\left(\frac{n\pi}{b}y\right) \tag{8.38}$$

Substitution of (8.34) into (8.30) gives the dispersion relation for a rectangular waveguide filled with isotropic medium as

$$\omega^2\mu\varepsilon - k^2 = \left(\frac{m\pi}{a}\right)^2 + \left(\frac{n\pi}{b}\right)^2$$

or

$$k^2 = \omega^2\mu\varepsilon - \left[\left(\frac{m\pi}{a}\right)^2 + \left(\frac{n\pi}{b}\right)^2\right] \tag{8.39}$$

Hence,

$$k = \sqrt{\omega^2\mu\varepsilon - \left[\left(\frac{m\pi}{a}\right)^2 + \left(\frac{n\pi}{b}\right)^2\right]} \tag{8.40}$$

Cut-off occurs when

$$\omega^2\mu\varepsilon - \left[\left(\frac{m\pi}{a}\right)^2 + \left(\frac{n\pi}{b}\right)^2\right] = 0 \tag{8.41}$$

The cut-off frequency, f_c, is found from (8.41) as

$$f_c = \frac{1}{2\pi\sqrt{\mu\varepsilon}}\sqrt{\left[\left(\frac{m\pi}{a}\right)^2 + \left(\frac{n\pi}{b}\right)^2\right]} \tag{8.42}$$

The summary of the field components, cut-off frequency and dispersion relation for rectangular waveguide filled with isotropic dielectric medium in Fig. 8.1 when the wave is propagating in y-direction are given in Table 8.1.

8.1.2 Waveguide Design with Gyrotropic Media

Now, assume that the rectangular waveguide shown in Fig. 8.1 is filled with the gyrotopic media in the existence of an external applied magnetic field, H_o, in y-direction. This is illustrated in Fig. 8.2.

Table 8.1 Field components, cut-off frequency, dispersion relation for rectangular waveguide filled with isotropic dielectric filling

Mode	Dispersion relation	f_c – cut-off freq	E field and H fields
TE_{mn}	$k = \sqrt{\omega^2 \mu \varepsilon - \left[\left(\frac{m\pi}{a}\right)^2 + \left(\frac{n\pi}{b}\right)^2\right]}$	$\frac{1}{2\pi\sqrt{\mu\varepsilon}} \sqrt{\left[\left(\frac{m\pi}{a}\right)^2 + \left(\frac{n\pi}{b}\right)^2\right]}$	$H_x(x,y) = A \frac{-ik}{(\omega^2 \mu\varepsilon - k^2)} \frac{m\pi}{a} \sin\left(\frac{m\pi}{a}x\right) \cos\left(\frac{n\pi}{b}y\right)$ $H_y(x,y) = A \frac{-ik}{(\omega^2 \mu\varepsilon - k^2)} \frac{n\pi}{b} \cos\left(\frac{m\pi}{a}x\right) \sin\left(\frac{n\pi}{b}y\right) pc$ $H_z(x,y) = A \cos\left(\frac{m\pi}{a}x\right) \cos\left(\frac{n\pi}{b}y\right)$ $E_x(x,y) = A \frac{-i\omega\mu}{(\omega^2 \mu\varepsilon - k^2)} \frac{n\pi}{b} \cos\left(\frac{m\pi}{a}x\right) \sin\left(\frac{n\pi}{b}y\right)$ $E_y(x,y) = A \frac{i\omega\mu}{(\omega^2 \mu\varepsilon - k^2)} \frac{m\pi}{a} \sin\left(\frac{m\pi}{a}x\right) \cos\left(\frac{n\pi}{b}y\right)$

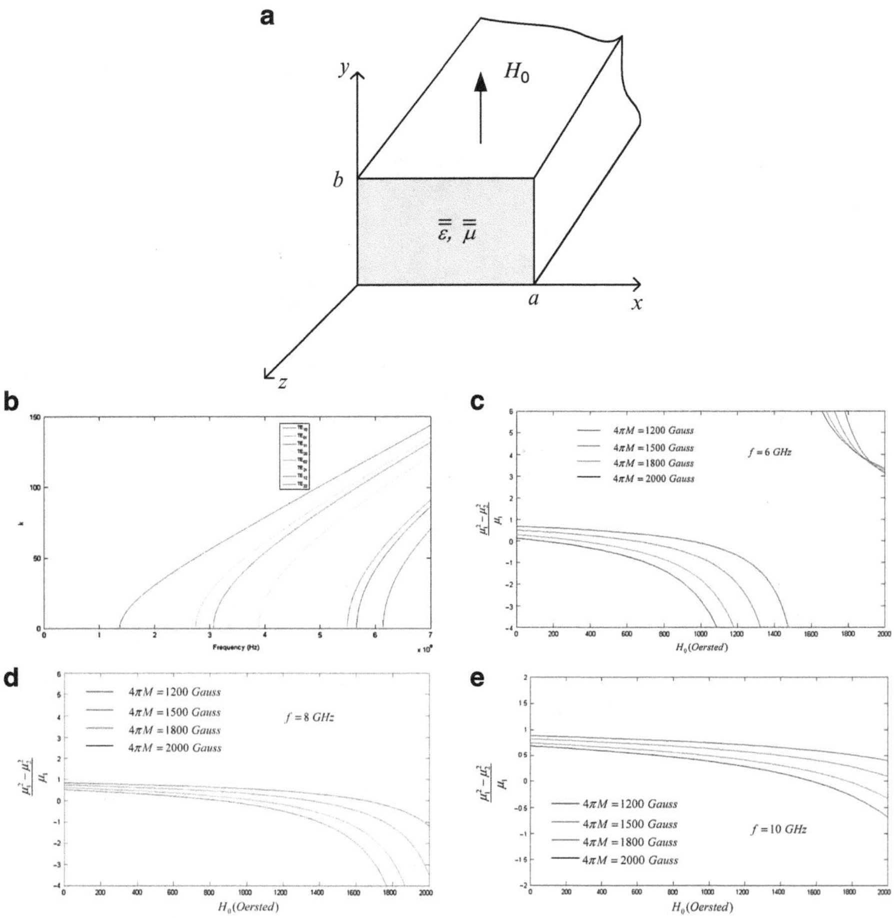

Fig. 8.2 (continued)

The permittivity and permeability tensors of the gyrotropic medium are given in Chap. 3 as

$$\bar{\bar{\varepsilon}} = \varepsilon_1(\bar{\bar{I}} - \hat{b}_0\hat{b}_0) + i\varepsilon_2(\hat{b}_0 \times \bar{\bar{I}}) + \varepsilon_3\hat{b}_0\hat{b}_0 \tag{8.43}$$

$$\bar{\bar{\mu}} = \mu_1(\bar{\bar{I}} - \hat{b}_0\hat{b}_0) + i\mu_2(\hat{b}_0 \times \bar{\bar{I}}) + \mu_3\hat{b}_0\hat{b}_0 \tag{8.44}$$

where $\hat{b}_0$ shows the direction of the external applied magnetic field. When the external applied magnetic field is in y-direction, $\hat{b}_0$ can be written as $\hat{b}_0 = [0 \quad 1 \quad 0]$. Then, (8.43) and (8.44) take the following matrix forms

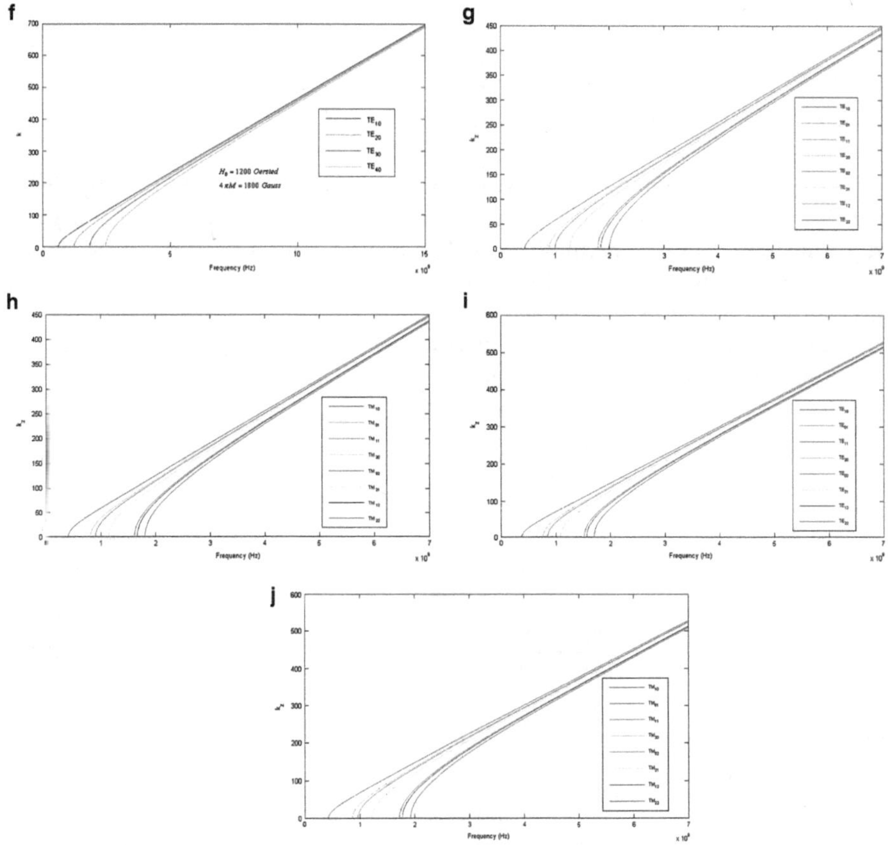

Fig. 8.2 (**a**) Geometry of the rectangular waveguide filled with transversely magnetized gyrotopic media. (**b**) Frequency response of the propagation constant for TE_{mn} modes. (**c**) Permeability parameters versus magnetic field intensity for various saturated magnetization levels at f = 6 GHz. (**d**) Permeability parameters versus magnetic field intensity for various saturated magnetization levels at f = 8 GHz. (**e**) Permeability parameters versus magnetic field intensity for various saturated magnetization levels at f = 10 GHz. (**f**) Frequency response of the propagation constant for TE_{mn} modes for a rectangular waveguide filled with magnetically gyrotropic medium. (**g**) Frequency response of the propagation constant for TE_{mn} modes for sapphire. (**h**) Frequency response of the propagation constant for TM_{mn} modes for sapphire. (**i**) Frequency response of the propagation constant for TE_{mn} modes for ceramic impregnated teflon. (**j**) Frequency response of the propagation constant for TM_{mn} modes for ceramic impregnated teflon

$$\bar{\bar{\varepsilon}} = \begin{bmatrix} \varepsilon_1 & 0 & -i\varepsilon_2 \\ 0 & \varepsilon_3 & 0 \\ i\varepsilon_2 & 0 & \varepsilon_1 \end{bmatrix} \tag{8.45}$$

and

$$\bar{\bar{\mu}} = \begin{bmatrix} \mu_1 & 0 & -i\mu_2 \\ 0 & \mu_3 & 0 \\ i\mu_2 & 0 & \mu_1 \end{bmatrix} \tag{8.46}$$

The elements of the permittivity and permeability tensors are given by (3.13) and (3.14), respectively. The waveguide analysis will be given for general gyrotropic medium using the permittivity and permeability tensors given by (8.45) and (8.46). The results can be simplified for gyromagnetic or gyroelectric medium by just replacing one of the tensors with a unit tensor.

When the permeability and permittivity tensors given by (8.45) and (8.46) are substituted into the wave equations given by (8.3)–(8.8), we obtain

$$\frac{\partial E_z}{\partial y} - \frac{\partial E_y}{\partial z} = i\omega\mu_1 H_x - \omega\mu_2 H_z \tag{8.47}$$

$$\frac{\partial E_x}{\partial z} - \frac{\partial E_z}{\partial x} = i\omega\mu_3 H_y \tag{8.48}$$

$$\frac{\partial E_y}{\partial x} - \frac{\partial E_x}{\partial y} = \omega\mu_2 H_x + i\omega\mu_1 H_z \tag{8.49}$$

and

$$\frac{\partial H_z}{\partial y} - \frac{\partial H_y}{\partial z} = -i\omega\varepsilon_1 E_x + \omega\varepsilon_2 E_z \tag{8.50}$$

$$\frac{\partial H_x}{\partial z} - \frac{\partial H_z}{\partial x} = -i\omega\varepsilon_3 E_y \tag{8.51}$$

$$\frac{\partial H_y}{\partial x} - \frac{\partial H_x}{\partial y} = -\omega\varepsilon_2 E_x - i\omega\varepsilon_1 E_z \tag{8.52}$$

The wave equations can be put in the following forms assuming that the wave is propagating in z-direction with the form that is given by (8.9),

$$\frac{\partial E_z}{\partial y} - ikE_y = i\omega\mu_1 H_x - \omega\mu_2 H_z \tag{8.53}$$

$$ikE_x - \frac{\partial E_z}{\partial x} = i\omega\mu_3 H_y \tag{8.54}$$

$$\frac{\partial E_y}{\partial x} - \frac{\partial E_x}{\partial y} = \omega\mu_2 H_x + i\omega\mu_1 H_z \tag{8.55}$$

and

$$\frac{\partial H_z}{\partial y} - ikH_y = -i\omega\varepsilon_1 E_x + \omega\varepsilon_2 E_z \tag{8.56}$$

$$ikH_x - \frac{\partial H_z}{\partial x} = -i\omega\varepsilon_3 E_y \tag{8.57}$$

$$\frac{\partial H_y}{\partial x} - \frac{\partial H_x}{\partial y} = -\omega\varepsilon_2 E_x - i\omega\varepsilon_1 E_z \tag{8.58}$$

The condition of no spatial dependence of the RF fields in the direction of the external applied magnetic fields, i.e. $\frac{\partial}{\partial y} = 0$, leads to separation of the waves into TE_{m0} and TM_{m0} modes for the rectangular waveguide when filled with the transversely magnetized gyrotropic medium.

8.1.2.1 TE_{m0} Modes

When the condition $\frac{\partial}{\partial y} = 0$ is applied for the wave equations given by (8.53)–(8.58), we obtain

$$-ikE_y = i\omega\mu_1 H_x - \omega\mu_2 H_z \tag{8.59}$$

$$\frac{\partial E_y}{\partial x} = \omega\mu_2 H_x + i\omega\mu_1 H_z \tag{8.60}$$

$$-i\omega\varepsilon_3 E_y = ikH_x - \frac{\partial H_z}{\partial x} \tag{8.61}$$

The field components H_x and H_z can be written in terms of E_y as

$$H_x = \frac{\mu_1}{\omega\left(\mu_1^2 - \mu_2^2\right)}\left(kE_y + \frac{\mu_2}{\mu_1}\frac{dE_y}{dx}\right) \tag{8.62}$$

$$H_z = -\frac{i\mu_1}{\omega\left(\mu_1^2 - \mu_2^2\right)}\left(k\frac{\mu_2}{\mu_1}E_y + \frac{dE_y}{dx}\right) \tag{8.63}$$

Substitution of (8.62) and (8.63) into (8.61) gives

Table 8.2 Field components, cut-off frequency, dispersion relation for rectangular waveguide filled with transversely magnetized gyrotropic filling

Mode	Dispersion relation	f_c – cut-off freq	E fields and H fields
$\mathrm{TE_{m0}}$	$k=\sqrt{\omega^2\left(\frac{\mu_1^2-\mu_2^2}{\mu_1}\right)\varepsilon_3-\left(\frac{m\pi}{a}\right)^2}$	$\frac{1}{2\pi\sqrt{\left(\frac{\mu_1^2-\mu_2^2}{\mu_1}\right)\varepsilon_3}}\left(\frac{m\pi}{a}\right)$	$E_z=A\sin\left(\frac{m\pi}{a}x\right)$ $H_x=\frac{A\mu_1}{\omega\left(\mu_1^2-\mu_2^2\right)}\left(\beta\sin\left(\frac{m\pi}{a}x\right)+\frac{\mu_2\left(\frac{m\pi}{a}\right)}{\mu_1}\cos\left(\frac{m\pi}{a}x\right)\right)$ $H_z=-\frac{iA\mu_1}{\omega\left(\mu_1^2-\mu_2^2\right)}\left(k\frac{\mu_2}{\mu_1}\sin\left(\frac{m\pi}{a}x\right)+\left(\frac{m\pi}{a}\right)\cos\left(\frac{m\pi}{a}x\right)\right)$

$$\frac{d^2E_y}{dx^2} + E_y\left(\omega^2\left(\frac{\mu_1^2 - \mu_2^2}{\mu_1}\right)\varepsilon_3 - k^2\right) = 0 \quad (8.64)$$

or

$$\frac{d^2E_y}{dx^2} + E_y k_\perp^2 = 0 \quad (8.65)$$

where

$$k_\perp^2 = \omega^2\left(\frac{\mu_1^2 - \mu_2^2}{\mu_1}\right)\varepsilon_3 - k^2 \quad (8.66)$$

Solution of the wave equation in (8.64) and (8.65) is

$$E_y = A\sin(k_\perp x) \quad (8.67)$$

Applying the boundary condition at x $= 0$, $E_y = 0$ gives

$$k_\perp = \frac{m\pi}{a} \text{ for } m = 1, 2, 3, ... \quad (8.68)$$

Hence, the dispersion relation for the rectangular waveguide filled with a transversely magnetized gyrotropic medium can be obtained from (8.66) and (8.68) as

$$k = \sqrt{\omega^2\left(\frac{\mu_1^2 - \mu_2^2}{\mu_1}\right)\varepsilon_3 - k_\perp^2} = \sqrt{\omega^2\left(\frac{\mu_1^2 - \mu_2^2}{\mu_1}\right)\varepsilon_3 - \left(\frac{m\pi}{a}\right)^2} \quad (8.69)$$

The cut-off frequency is calculated from

$$\omega^2\left(\frac{\mu_1^2 - \mu_2^2}{\mu_1}\right)\varepsilon_3 - \left(\frac{m\pi}{a}\right)^2 = 0 \quad (8.70)$$

So,

$$f_c = \frac{1}{2\pi\sqrt{\left(\frac{\mu_1^2 - \mu_2^2}{\mu_1}\right)\varepsilon_3}}\left(\frac{m\pi}{a}\right) \quad (8.71)$$

The magnetic field components are found by substituting (8.67) into (8.62) and (8.63) as

$$H_x = \frac{A\mu_1}{\omega\left(\mu_1^2 - \mu_2^2\right)} \left(k \sin\left(\frac{m\pi}{a}x\right) + \frac{\mu_2 \left(\frac{m\pi}{a}\right)}{\mu_1} \cos\left(\frac{m\pi}{a}x\right) \right) \tag{8.72}$$

$$H_z = -\frac{iA\mu_1}{\omega\left(\mu_1^2 - \mu_2^2\right)} \left(k\frac{\mu_2}{\mu_1} \sin\left(\frac{m\pi}{a}x\right) + \left(\frac{m\pi}{a}\right) \cos\left(\frac{m\pi}{a}x\right) \right) \tag{8.73}$$

The table illustrating the results for the rectangular waveguide filled with gyrotropic medium is given in Table 8.2.

8.1.3 Waveguide Design with Anisotropic Media

In this section, we assume the dielectric filling for the rectangular waveguide illustrated in Fig. 8.1 is a uniaxailly anisotropic medium with the following permittivity and permeability tensors

$$\bar{\bar{\varepsilon}} = \begin{bmatrix} \varepsilon_{11} & 0 & 0 \\ 0 & \varepsilon_{11} & 0 \\ 0 & 0 & \varepsilon_{33} \end{bmatrix}, \quad \bar{\bar{\mu}} = \mu_0 \bar{\bar{I}} \tag{8.74}$$

The optic axis of the anisotropic medium is assumed to be oriented in z-direction. The dispersion relation for a vertically uniaxial anisotropic medium is obtained using (2.17) as

$$\bar{\bar{W}}_E \cdot \bar{E} = 0 \tag{8.75}$$

where

$$\bar{\bar{W}}_E = \left[\bar{\bar{k}} \cdot \bar{\bar{\mu}}^{-1} \cdot \bar{\bar{k}} + k_0^2 \bar{\bar{\varepsilon}} \right] \tag{8.76}$$

The characteristic values for ordinary and extraordinary waves are obtained when (8.74) is substituted into (8.75).

$$k_{\rho I}^2 = k_0^2 \varepsilon_{11} - k_z^2 \tag{8.77}$$

$$k_{\rho II}^2 = k_0^2 \varepsilon_{33} - k_z^2 \frac{\varepsilon_{33}}{\varepsilon_{11}} \tag{8.78}$$

where

$$k^2 = k_\rho^2 + k_z^2 = k_x^2 + k_y^2 + k_z^2$$

When (8.77) and (8.78) are substituted into (8.75), we obtain the corresponding characteristic field vectors for the characteristic values given by (8.77) and (8.78). The characteristic field vectors for ordinary and extraordinary wave are

$$\bar{E}_1 = \hat{x}A\cos\left(\tan^{-1}\left(\frac{k_y}{k_x}\right)\right) + \hat{y}B\sin\left(\tan^{-1}\left(\frac{k_y}{k_x}\right)\right) \tag{8.79}$$

$$\bar{E}_2 = \frac{\varepsilon_{33}k_z}{\varepsilon_{11}\sqrt{\varepsilon_{33}k_0^2 - k_z^2\frac{\varepsilon_{33}}{\varepsilon_{11}}}} \times \left[\hat{x}\cos\left(\tan^{-1}\left(\frac{k_y}{k_x}\right)\right)E_{2z} + \hat{y}B\sin\left(\tan^{-1}\left(\frac{k_y}{k_x}\right)\right)E_{2z}\right] + \hat{z}E_{2z} \tag{8.80}$$

It is clear from (8.79) and (8.80) that ordinary wave, $\bar{E}_1$, represents TE waves and extraordinary wave or $\bar{E}_2$ represent TM waves. Application of boundary conditions gives

$$k_{x1} = \frac{m_1\pi}{a} \text{ for } m_1 = 1, 2, 3, \ldots \tag{8.81}$$

$$k_{y1} = \frac{n_1\pi}{b} \text{ for } \mathrm{n}_1 = 0, 1, 2, 3, \ldots \tag{8.82}$$

for TE or ordinary waves and

Table 8.3 Cut-off frequency, dispersion relation for rectangular waveguide filled with vertically uniaxial anisotropic filling

Mode	Dispersion relation	f_c – cut-off frequency
TE_{mn}	$k_z = \sqrt{k_0^2\varepsilon_{11} - \left[\left(\frac{m_1\pi}{a}\right)^2 + \left(\frac{n_1\pi}{b}\right)^2\right]}$	$f_c = \frac{1}{2\pi\sqrt{\mu_0\varepsilon_{11}}}\sqrt{\left[\left(\frac{m\pi}{a}\right)^2 + \left(\frac{n\pi}{b}\right)^2\right]}$
TM_{mn}	$k_z = \sqrt{k_0^2\varepsilon_{11} - \frac{\varepsilon_{11}}{\varepsilon_{33}}\left[\left(\frac{m_2\pi}{a}\right)^2 + \left(\frac{n_2\pi}{b}\right)^2\right]}$	$f_c = \frac{1}{2\pi\sqrt{\mu_0\varepsilon_{33}}}\sqrt{\left[\left(\frac{m\pi}{a}\right)^2 + \left(\frac{n\pi}{b}\right)^2\right]}$

Table 8.4 Standard rectangular waveguide dimensions

Waveguide standard	Inside dimension – a (in)	Inside dimension – b (in)
WR-2300	23	11.5
WR-2100	21	10.5
WR-1800	18	9
WR-1500	15	7.5
WR-1150	11.5	5.75
WR-1000	9.975	4.875
WR-770	7.7	3.385
WR-650	6.5	3.25
WR-430	4.3	2.15

Table 8.5 TE_{mn} modes for WR-430

Waveguide standard	TE_{mn} mode	Calculated cut-off freq, f_c (GHz)
WR-430	TE_{10}	1.3727
	TE_{01}	2.7455
	TE_{11}	3.0696
	TE_{20}	2.7455
	TE_{02}	5.4910
	TE_{21}	3.8827
	TE_{12}	5.6600
	TE_{22}	6.1391

$$k_{x2} = \frac{m_2\pi}{a} \text{ for } m_2 = 1, 2, 3, \ldots \tag{8.83}$$

$$k_{y2} = \frac{n_2\pi}{b} \text{ for } \mathrm{n}_2 = 0, 1, 2, 3, \ldots \tag{8.84}$$

for TM or extraordinary waves. As a result, the substitution of (8.81) and (8.82) into (8.77) gives the dispersion relation for TE waves as

$$k_{\rho I}^2 = k_0^2\varepsilon_{11} - k_z^2 = \left(\frac{m_1\pi}{a}\right)^2 + \left(\frac{n_1\pi}{b}\right)^2 \tag{8.85}$$

or

$$k_z = \sqrt{k_0^2\varepsilon_{11} - \left[\left(\frac{m_1\pi}{a}\right)^2 + \left(\frac{n_1\pi}{b}\right)^2\right]} \tag{8.86}$$

The cut-off frequency for TE waves is

$$f_c = \frac{1}{2\pi\sqrt{\mu_0\varepsilon_{11}}}\sqrt{\left[\left(\frac{m\pi}{a}\right)^2 + \left(\frac{n\pi}{b}\right)^2\right]} \tag{8.87}$$

Similarly, substitution of (8.83) and (8.84) into (8.78) gives the dispersion relation for TM waves as

$$k_{\rho II}^2 = k_0^2\varepsilon_{33} - k_z^2\frac{\varepsilon_{33}}{\varepsilon_{11}} = \left(\frac{m_2\pi}{a}\right)^2 + \left(\frac{n_2\pi}{b}\right)^2 \tag{8.88}$$

or

$$k_z = \sqrt{k_0^2\varepsilon_{11} - \frac{\varepsilon_{11}}{\varepsilon_{33}}\left[\left(\frac{m_2\pi}{a}\right)^2 + \left(\frac{n_2\pi}{b}\right)^2\right]} \tag{8.89}$$

Table 8.6 TE_{mn} modes for WR-430 waveguide filled with a magnetically gyrotropic medium for $4\pi M = 1800$ Gauss

Saturated magnetization	H_0 (Oersted)	TE_{mn} mode	Calculated cut-off freq, f_c (GHz)
$4\pi M = 1800$ Gauss	600	TE_{10}	0.5040
$f = 10GHz$		TE_{20}	1.0080
		TE_{30}	1.5120
		TE_{40}	2.0160
	800	TE_{10}	0.5288
		TE_{20}	1.0576
		TE_{30}	1.5864
		TE_{40}	2.1152
	1,000	TE_{10}	0.5639
		TE_{20}	1.1279
		TE_{30}	1.6919
		TE_{40}	2.2559
	1,200	TE_{10}	0.6187
		TE_{20}	1.2375
		TE_{30}	1.8563
		TE_{40}	2.4750

Table 8.7 TE_{mn} modes for WR-430 waveguide filled with a magnetically gyrotropic medium for $4\pi M = 2200$ Gauss

Saturated magnetization	H_0 (Oersted)	TE_{mn} mode	Calculated cut-off freq, f_c (GHz)
$4\pi M = 2200$ Gauss	600	TE_{10}	0.594
$f = 10GHz$		TE_{20}	1.189
		TE_{30}	1.784
		TE_{40}	2.379
	800	TE_{10}	0.658
		TE_{20}	1.316
		TE_{30}	1.974
		TE_{40}	2.632
	1,000	TE_{10}	0.772
		TE_{20}	1.544
		TE_{30}	2.317
		TE_{40}	3.089
	1,200	TE_{10}	1.067
		TE_{20}	2.135
		TE_{30}	3.203
		TE_{40}	4.270

The cut-off frequency for TM waves is

$$f_c = \frac{1}{2\pi\sqrt{\mu_0 \varepsilon_{33}}} \sqrt{\left[\left(\frac{m\pi}{a}\right)^2 + \left(\frac{n\pi}{b}\right)^2 \right]} \tag{8.90}$$

The table giving all the design parameters for a rectangular waveguide filled with uniaxially anisotropic medium is illustrated in Table 8.3.

Table 8.8 TE_{mn} and TM_{mn} modes for WR-430 waveguide filled with positively uniaxial anisotropic medium, sapphire

Waveguide standard	TE_{mn} mode	Calculated cut-off freq, f_c (GHz)	TM_{mn} mode	Calculated cut-off freq, f_c (GHz)
WR-430	TE_{10}	0.447	TM_{10}	0.403
	TE_{01}	0.8954	TM_{01}	0.806
	TE_{11}	1.001	TM_{11}	0.901
	TE_{20}	0.895	TM_{20}	0.806
	TE_{02}	1.791	TM_{02}	1.612
	TE_{21}	1.266	TM_{21}	1.14
	TE_{12}	1.846	TM_{12}	1.662
	TE_{22}	2.002	TM_{22}	1.802

Table 8.9 TE_{mn} and TM_{mn} modes for WR-430 waveguide filled with negatively uniaxial anisotropic medium, ceramic impregnated teflon

Waveguide standard	TE_{mn} mode	Calculated cut-off freq, f_c (GHz)	TM_{mn} mode	Calculated cut-off freq, f_c (GHz)
WR-430	TE_{10}	0.380	TM_{10}	0.429
	TE_{01}	0.761	TM_{01}	0.859
	TE_{11}	0.851	TM_{11}	0.961
	TE_{20}	0.761	TM_{20}	0.859
	TE_{02}	1.522	TM_{02}	1.719
	TE_{21}	1.076	TM_{21}	1.215
	TE_{12}	1.569	TM_{12}	1.772
	TE_{22}	1.702	TM_{22}	1.922

8.1.4 Design Examples

Some of the standard rectangular waveguide dimensions used in RF applications are given in Table 8.4 below.

8.1.4.1 Isotropic Case

The calculated cut-off frequencies of the TE_{mn} modes for the rectangular waveguide WR-430 when filled with air are given in Table 8.5.

The frequency response of the propagation constant using (8.40) is illustrated in Fig. 8.2b. It is obvious that the lowest cut-off frequency occurs for TE_{10} mode. The cut-off frequencies in Fig. 8.2b match the calculated cut-off frequencies given in Table 8.5 for all the propagating modes.

8.1.4.2 Gyrotropic Case

In this case, the standard rectangular waveguide, WR-430, is filled with a magnetically gyrotropic medium, i.e. ferrite. The dispersion relation given by (8.69) is

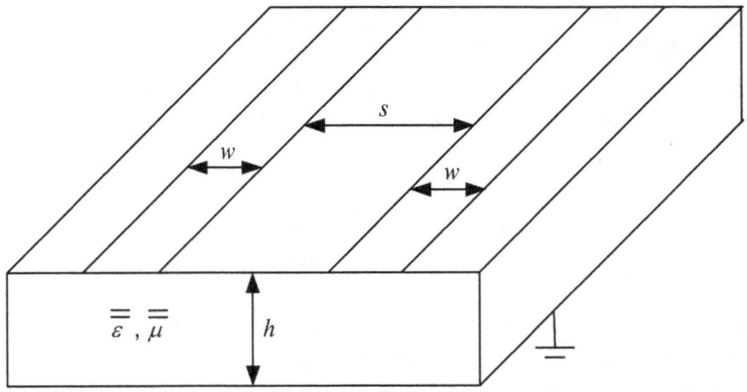

Fig. 8.3 Geometry of the symmetrical coupler

function of the term $\frac{\mu_1^2-\mu_2^2}{\mu_1}$. This term is also referred as the effective permeability constant of the ferrite material. The response of this term versus external magnetic field intensity is given in Fig. 8.2c–e for various saturated magnetization levels at 6 GHz, 8 GHz and 10 GHz, respectively. The values of $\frac{\mu_1^2-\mu_2^2}{\mu_1}$ can be found from the graph provided for the given saturated magnetization level, magnetic field intensity and operational frequency. It can then be used in (8.69) to calculate the frequency response of the propagation constant, k, and the corresponding cut-off frequency.

At this point, we can obtain the frequency response of the propagation constant with the design curves given in Fig. 8.2c–e. The ferrite material which has $\varepsilon_r = 12$ is used as a filling for the rectangular waveguide, WR-430, and biased with $H_0 = 1200\ Oersted$ at $f = 8GHz$ with 1,800 Gauss saturated magnetization level. The frequency response of k in the frequency range of 0–15 GHz for four TE modes are obtained and illustrated in Fig. 8.2f. The calculated cut-off frequencies for the TE modes for various values of the magnetic field intensities when saturated magnetization levels are 1,800 Gauss and 2,200 Gauss and operational frequency is 10 GHz are tabulated in Tables 8.6 and 8.7.

8.1.4.3 Anisotropic Case

The calculated cut-off frequencies for the ordinary waves, TE modes, and extraordinary waves, TM modes, for the standard rectangular waveguide, WR-430, when filled with positively uniaxial anisotropic medium, sapphire, are given in Table 8.8.

The permittivity tensor of the sapphire is given as

$$\bar{\bar{\varepsilon}} = \begin{bmatrix} 9.4 & 0 & 0 \\ 0 & 9.4 & 0 \\ 0 & 0 & 11.6 \end{bmatrix}$$

The frequency response of the propagation constant using (8.86) and (8.89) are given in Fig. 8.2g, h for TE and TM modes, respectively.

Now, we change the rectangular waveguide filling to negatively uniaxial anisotropic medium, ceramic impregnated teflon, with the following permittivity tensor

$$\bar{\bar{\varepsilon}} = \begin{bmatrix} 13 & 0 & 0 \\ 0 & 13 & 0 \\ 0 & 0 & 10.2 \end{bmatrix}$$

The calculated cut-off frequencies for the ordinary waves, TE modes, and extraordinary waves, TM modes, for the standard rectangular waveguide, WR-430, when filled with negatively uniaxial anisotropic medium, ceramic impregnated teflon, are given in Table 8.9.

The frequency response of the propagation constant using (8.86) and (8.89) are given in Fig. 8.2i, j for TE and TM modes, for WR-430 when filled with ceramic impregnated teflon. The numerical and analytical values of the cut-off frequencies given in Tables and Figures agree as illustrated.

8.2 Microstrip Directional Coupler Design

Symmetrical coupled line structures are commonly used as a directional coupler to detect signal at the frequency of interest. The geometry of symmetrical structure is illustrated in Fig. 8.3.

In this section, we give the detailed analysis of the coupled structures which are used as directional couplers using isotropic, anisotropic, and gyrotropic media.

8.2.1 Microstrip Directional Coupler Design Using Isotropic Medium

Design equations for the symmetrical coupler shown in Fig. 8.3 when the medium is isotropic with the following parameters

$$\bar{\bar{\varepsilon}} = \varepsilon \bar{\bar{I}} \quad \text{and} \quad \bar{\bar{\mu}} = \mu \bar{\bar{I}} \tag{8.91}$$

can be obtained using the quasi-static analysis is given in [2]. The design procedure begins with the identification of the coupling level, port impedances, and operational frequency. Once these three design parameters are known, the required physical coupler dimensions can be calculated. The even and odd impedances of the coupled lines are determined from

$$Z_{oe} = Z_0 \sqrt{\frac{1 + 10^{C/20}}{1 - 10^{C/20}}} \tag{8.92}$$

$$Z_{oo} = Z_0\sqrt{\frac{1 - 10^{C/20}}{1 + 10^{C/20}}} \tag{8.93}$$

where C is the forward coupling level in dB. The spacing between the coupled lines are found using

$$s/h = \frac{2}{\pi}\cosh^{-1}\left[\frac{\cosh\left[\frac{\pi}{2}\left(\frac{w}{h}\right)'_{so}\right] + \cosh\left[\frac{\pi}{2}\left(\frac{w}{h}\right)_{so}\right] - 2}{\cosh\left[\frac{\pi}{2}\left(\frac{w}{h}\right)'_{so}\right] - \cosh\left[\frac{\pi}{2}\left(\frac{w}{h}\right)_{se}\right]}\right] \tag{8.94}$$

$(w/h)_{se}$ and $(w/h)_{so}$ are the shape ratios for the equivalent single case corresponding to even-mode and odd-mode geometry, respectively. $(w/h)'_{so}$ is

$$\left(\frac{W}{h}\right)'_{so} = 0.78\left(\frac{W}{h}\right)_{so} + 0.1\left(\frac{W}{h}\right)_{se} \tag{8.95}$$

(w/h) is the shape ratio for the single microstrip line and it is calculated from

$$\frac{w}{h} = \frac{8\sqrt{\left[\exp\left(\frac{R}{42.4}\sqrt{(\varepsilon_r + 1)}\right) - 1\right]\frac{7+(4/\varepsilon_r)}{11} + \frac{1+(1/\varepsilon_r)}{0.81}}}{\left[\exp\left(\frac{R}{42.4}\sqrt{\varepsilon_r + 1}\right) - 1\right]} \tag{8.96}$$

where

$$R = \frac{Z_{oe}}{2} \quad \text{or} \quad R = \frac{Z_{oo}}{2} \tag{8.97}$$

$(w/h)_{se}$ and $(w/h)_{so}$ are found from

$$(w/h)_{se} = (w/h)\Big|_{R = Z_{ose}} \tag{8.98}$$

and

$$(w/h)_{so} = (w/h)\Big|_{R = Z_{oso}} \tag{8.99}$$

Z_{ose} and Z_{oso} are the characteristic impedances corresponding to single microstrip shape ratios $(w/h)_{se}$ and $(w/h)_{so}$, respectively. They are given by the following relations.

$$Z_{ose} = \frac{Z_{oe}}{2} \tag{8.100}$$

$$Z_{oso} = \frac{Z_{oo}}{2} \tag{8.101}$$

After the spacing ratio s/h for the coupled lines is calculated, w/h for the coupled lines is found from

$$\left(\frac{w}{h}\right) = \frac{1}{\pi}\cosh^{-1}(d) - \frac{1}{2}\left(\frac{s}{h}\right) \tag{8.102}$$

where

$$d = \frac{\cosh\left[\frac{\pi}{2}\left(\frac{w}{h}\right)_{se}\right](g+1) + g - 1}{2} \tag{8.103}$$

$$g = \cosh\left[\frac{\pi}{2}\left(\frac{s}{h}\right)\right] \tag{8.104}$$

Finally, the physical length of the directional coupler is obtained using

$$l = \frac{\lambda}{4} = \frac{c}{4f\sqrt{\varepsilon_{eff}}} \tag{8.105}$$

where $c = 3 * 10^8\ m/\sec$, f is operational frequency in Hz, and ε_{eff} is the effective permittivity constant [3]. ε_{eff} is determined from the following relation.

$$\varepsilon_{eff} = \left[\frac{\sqrt{\varepsilon_{effe}} + \sqrt{\varepsilon_{effo}}}{2}\right]^2 \tag{8.106}$$

ε_{effe} and ε_{effo} are the effective permittivity constants of the coupled structure for odd and even modes, respectively. ε_{effe} and ε_{effo} depend on even and odd mode capacitances C_e and C_o as

$$\varepsilon_{effe} = \frac{C_e}{C_{e1}} \tag{8.107}$$

$$\varepsilon_{effo} = \frac{C_o}{C_{o1}} \tag{8.108}$$

The even mode capacitance C_e is

$$C_e = C_p + C_f + C_f^{'} \tag{8.109}$$

C_p is the parallel plate capacitance, C_f is the fringing capacitance when the microstrip is being taken alone as if it were a single strip. The equations defining C_p,C_f, and $C_f^{'}$ are given as

$$C_p = \varepsilon_0\varepsilon_r\frac{w}{h} \tag{8.110}$$

$$C_f = \frac{\sqrt{\varepsilon_{seff}}}{2cZ_0} - \frac{C_p}{2} \tag{8.111}$$

$$C_f' = \frac{C_f}{1 + A\left(\frac{h}{s}\right)\tanh\left(\frac{10h}{s}\right)}\left(\frac{\varepsilon_r}{\varepsilon_{seff}}\right)^{1/4} \quad (8.112)$$

where

$$\varepsilon_{seff} = \frac{\varepsilon_r + 1}{2} - \frac{\varepsilon_r - 1}{2}F(w/h) \quad (8.113)$$

and

$$F(w/h) = \begin{Bmatrix} (1 + 12h/w)^{-1/2} + 0.041(1 - w/h)^2 \text{ for } \left(\frac{w}{h} \le 1\right) \\ (1 + 12h/w)^{-1/2} \text{ for } \left(\frac{w}{h} \ge 1\right) \end{Bmatrix} \quad (8.114)$$

$$A = \exp\left[-0.1\exp\left(2.33 - 1.5\frac{w}{h}\right)\right] \quad (8.115)$$

The odd mode capacitance C_o is

$$C_o = C_p + C_f + C_{ga} + C_{gd} \quad (8.116)$$

C_{ga} is the capacitance term in odd mode for the fringing field across the gap in the air region and C_{gd} represents the capacitance in odd mode for the fringing field across the gap in the dielectric region. C_{ga} and C_{gd} are given as

$$C_{ga} = \varepsilon_0\frac{K(k')}{K(k)} \quad (8.117)$$

$$C_{gd} = \frac{\varepsilon_0\varepsilon_r}{\pi}\ln\left\{\coth\left(\frac{\pi}{4}\frac{s}{h}\right)\right\} + 0.65C_f\left[\frac{0.02}{\left(\frac{s}{h}\right)}\sqrt{\varepsilon_r} + \left(1 - \frac{1}{\varepsilon_r^2}\right)\right] \quad (8.118)$$

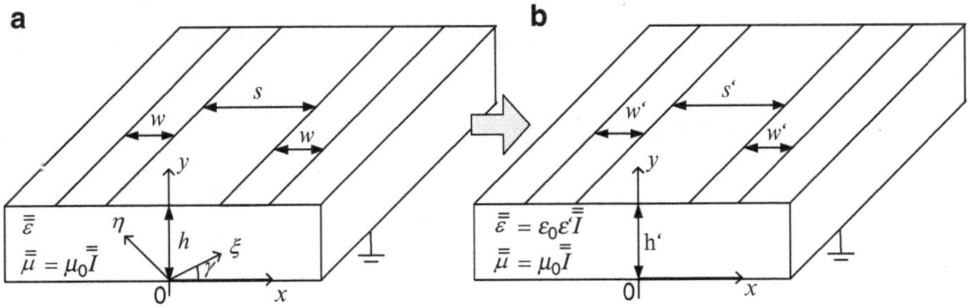

Fig. 8.4 (**a**) Geometry of the symmetrical coupler on anisotropic substrate. (**b**) Geometry of the symmetrical coupler after transformation to equivalent isotropic substrate

where

$$\frac{K(k')}{K(k)} = \left\{ \begin{array}{l} \frac{1}{\pi}\ln\left[2\frac{1+\sqrt{k'}}{1-\sqrt{k'}}\right], 0 \le \mathrm{k}^2 \le 0.5 \\ \frac{\pi}{\ln\left[2\frac{1+\sqrt{k'}}{1-\sqrt{k'}}\right]}, 0.5 \le \mathrm{k}^2 \le 1 \end{array} \right\} \tag{8.119}$$

$$k = \frac{\left(\frac{s}{h}\right)}{\left(\frac{s}{h}\right) + \left(\frac{2w}{h}\right)} \tag{8.120}$$

and

$$k' = \sqrt{1-k^2} \tag{8.121}$$

C_{e1} and C_{o1} are calculated using the odd and even mode impedances as

$$C_{e1} = \frac{1}{c^2 C_e Z_{oe}^2} \tag{8.122}$$

$$C_{o1} = \frac{1}{c^2 C_o Z_{oo}^2} \tag{8.123}$$

Equations given by (8.92)–(8.123) give the complete design equations to obtain the physical dimensions of the directional coupler when the medium is isotropic at the desired coupling level, operational frequency, and port impedances.

8.2.2 *Microstrip Directional Coupler Design Using Anisotropic Medium*

The directional coupler design using anisotropic materials have attracted engineers because of their benefits due to natural anisotropy. This feature can be used as a knob to adjust some of the critical design parameters. The most practical method

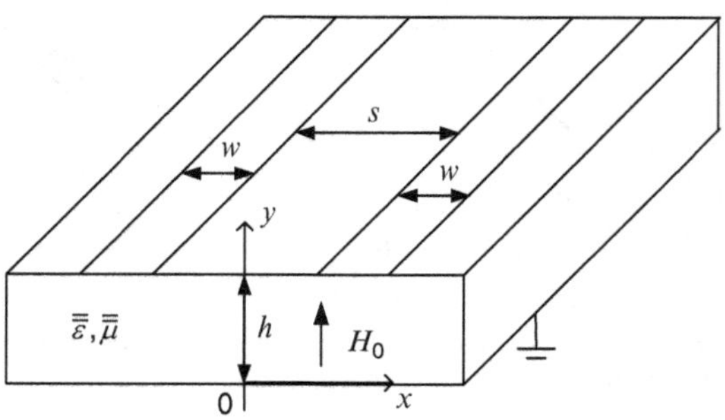

Fig. 8.5 Geometry of the symmetrical coupler on gyrotropic substrate

that can be used to design directional couplers on anisotropic substrates is given in [4]. The method simplifies anisotropic directional coupler design to isotropic directional coupler design with the transformations given by (8.124)–(8.128). The geometry of the coupler is illustrated in Fig. 8.4.

$$\varepsilon' = \sqrt{\varepsilon_\xi \varepsilon_\eta} \tag{8.124}$$

$$W' = W \tag{8.125}$$

$$h' = \frac{(\alpha h)}{[(\alpha^2 - 1)\cos^2(\gamma) + 1]} \tag{8.126}$$

$$s' = s \tag{8.127}$$

where

$$\alpha = \sqrt{\frac{\varepsilon_\eta}{\varepsilon_\xi}} \tag{8.128}$$

The permittivity tensor of anisotropic the medium is defined as

$$\bar{\bar{\varepsilon}} = \varepsilon_0 \begin{bmatrix} \varepsilon_\xi & 0 \\ 0 & \varepsilon_\eta \end{bmatrix} \tag{8.129}$$

in the $\xi - \eta$ plane. The permittivity tensor is obtained with rotating $x - y$ plane by an angle γ. In the configuration illustrated in Fig. 8.4(a), the principles axis of the anisotropic medium is defined by $\xi - \eta$. ε_ξ and ε_η are the relative permittivity constants of the anisotropic medium in the direction of principle axes. Once the transformation is accomplished using (8.124)–(8.128), the complete design equations given by (8.92)–(8.123) are used to obtain the final physical dimensions of the directional coupler with an anisotropic substrate.

8.2.3 *Microstrip Directional Coupler Design Using Gyrotropic Medium*

The non-reciprocity effect in gyrotropic materials is one of the critical material properties that make them superior to reciprocal counterparts. The analysis of the directional coupler is based on the Galerkin's method in the Fourier domain [5–7]. The geometry of coupler using gyrotopic medium is shown in Fig. 8.5.

When the applied external magnetic field is in y-direction, the permittivity and permeability tensors are given by

$$\bar{\bar{\varepsilon}} = \begin{bmatrix} \varepsilon_1 & 0 & -i\varepsilon_2 \\ 0 & \varepsilon_3 & 0 \\ i\varepsilon_2 & 0 & \varepsilon_1 \end{bmatrix} \tag{8.130}$$

and

$$\bar{\bar{\mu}} = \begin{bmatrix} \mu_1 & 0 & -i\mu_2 \\ 0 & \mu_3 & 0 \\ i\mu_2 & 0 & \mu_1 \end{bmatrix} \tag{8.131}$$

When the boundary conditions are applied for the coupler illustrated in Fig. 8.5., we obtain

$$\begin{bmatrix} \bar{E}_x \\ \bar{E}_z \end{bmatrix} = \begin{bmatrix} \bar{\bar{G}}_{xx} & \bar{\bar{G}}_{xz} \\ \bar{\bar{G}}_{zx} & \bar{\bar{G}}_{zz} \end{bmatrix} \begin{bmatrix} \bar{J}_x \\ \bar{J}_z \end{bmatrix} \tag{8.132}$$

In (8.132), $\bar{E}_x$,$\bar{E}_z$ are the Fourier transformed tangential electric field vectors and $\bar{J}_x$, $\bar{J}_z$ are the Fourier transformed electric current density components. We assume that the gyrotropic material in Fig. 8.5 is magnetically gyrotropic, i.e., ferrite, to simplify our analysis. Then, the permittivity tensor takes the following form

$$\bar{\bar{\varepsilon}} = \varepsilon\bar{\bar{I}} \tag{8.133}$$

The elements of the dyadic Green's functions in (8.132) are found and given by using (8.134)–(8.137) as

$$\bar{\bar{G}}_{xx} = -K\left[\chi_1 + \frac{\beta}{\alpha}\chi_2\right] \tag{8.134}$$

$$\bar{\bar{G}}_{xz} = K\left[\frac{\beta}{\alpha}\chi_1 + \chi_2\right] \tag{8.135}$$

$$\bar{\bar{G}}_{zx} = -K\left[\frac{\alpha}{\beta}\chi_1 - \chi_2\right] \tag{8.136}$$

$$\bar{\bar{G}}_{zz} = -K\left[-\chi_1 + \frac{\alpha}{\beta}\chi_2\right] \tag{8.137}$$

where

$$K = \frac{i\alpha\beta}{\omega\left(\alpha^2 + \beta^2\right)} \tag{8.138}$$

$$\chi_1 = \left(i\omega^2 \varepsilon \mu_2\right) \frac{\left(\frac{\coth(\gamma_{e1} h)}{\gamma_{e1}}\right)}{\left(\varepsilon_0 + \frac{\varepsilon \gamma_o}{\gamma_{e1}} \coth(\gamma_{e1} h)\right)\left(\frac{\mu_1}{\mu_3} + \frac{\gamma_{h1}}{\gamma_0} \coth(\gamma_{h1} d)\right)} \tag{8.139}$$

$$\chi_2 = \frac{(\omega^2 \mu_1)}{\gamma_0 \left(\frac{\mu_1}{\mu_3} + \frac{\gamma_{h1}}{\gamma_0} \coth(\gamma_{h1} d)\right)} \tag{8.140}$$

$$\gamma_{h1}^2 = \frac{\mu_1}{\mu_3}\left(\alpha^2 + \beta^2 - k_1^2\right) \tag{8.141}$$

$$\gamma_{e1}^2 = \left(\alpha^2 + \beta^2 - k_2^2\right) \tag{8.142}$$

$$\gamma_o^2 = \left(\alpha^2 + \beta^2 - k_o^2\right) \tag{8.143}$$

$$k_o^2 = \omega^2 \mu_0 \varepsilon_0 \tag{8.144}$$

$$k_1^2 = \omega^2 \mu_0 \varepsilon \tag{8.145}$$

$$k_2^2 = \omega^2 \left(\frac{\mu_1^2 - \mu_2^2}{\mu_1}\right) \varepsilon \tag{8.146}$$

Propagation constants are found with the Parseval's theorem and the application of Galerkin's method. Once the propagation constants are found, the physical dimensions of the coupler with ferrite substrate can be determined by calculating the odd and even mode impedances from

$$Z_o = \frac{2P}{I^2} \tag{8.147}$$

where

$$P = \frac{1}{2} \mathrm{Re}\left\{ \int\int \left(E_x H_y^* - E_y H_x^*\right) dxdy \right\} \tag{8.148}$$

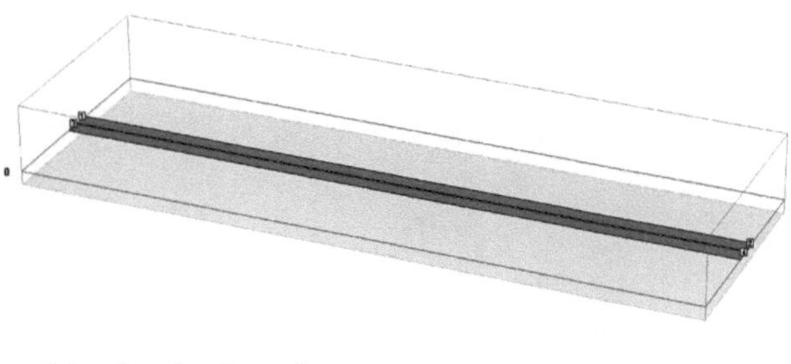

Fig. 8.6 Layout of the directional coupler

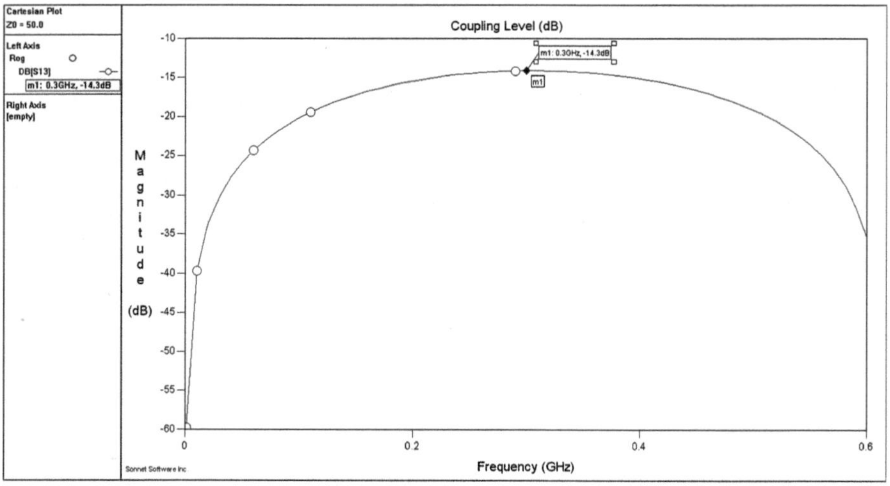

Fig. 8.7 Coupling level for symmetrical directional coupler using isotropic medium

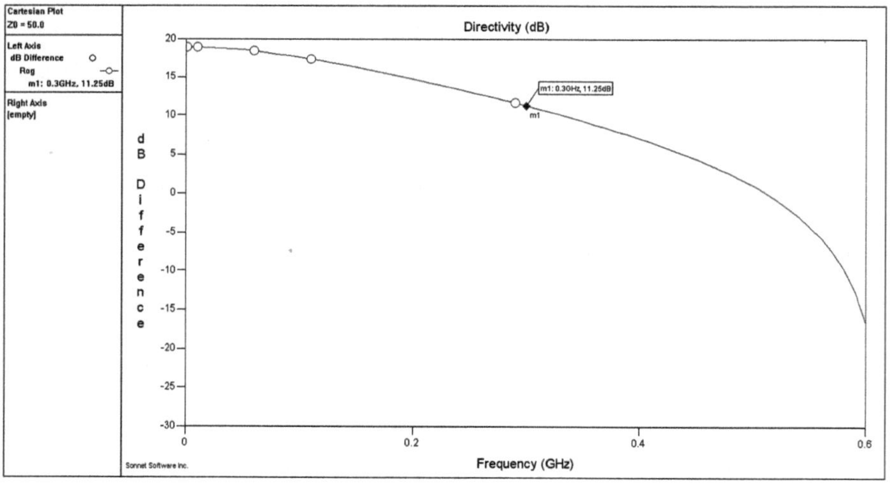

Fig. 8.8 Directivity level for symmetrical directional coupler using isotropic medium

and

$$I = \int_{-W/2}^{W/2} J_z(x)dx \tag{8.149}$$

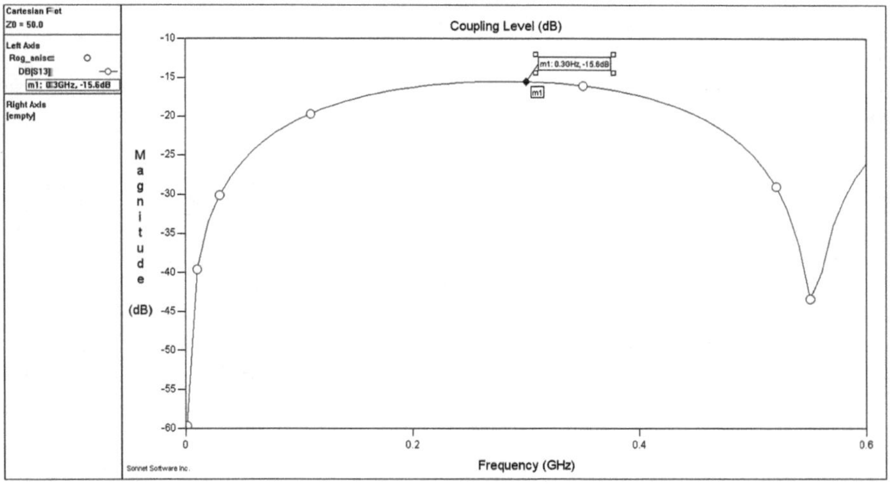

Fig. 8.9 Coupling level for symmetrical directional coupler using positively anisotropic medium

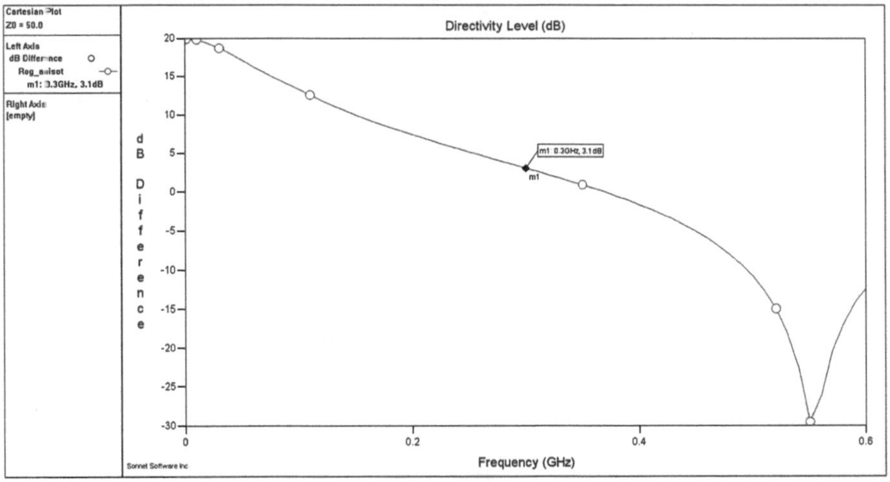

Fig. 8.10 Directivity level for symmetrical directional coupler using positively anisotropic medium

The desired coupling level is then found using (8.92) and (8.93) as described in Sect. 8.2.1.

8.2.4 Design Examples

In this section, we give design examples for the directional couplers using the analysis methods described in Sects. 8.2.1–8.2.3. The physical dimension of the symmetrical directional coupler for 15 dB coupling level is designed based on

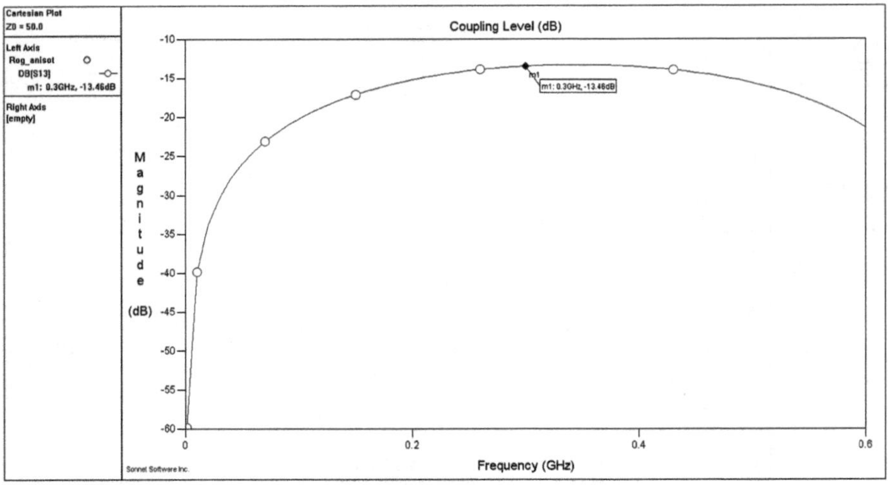

Fig. 8.11 Coupling level for symmetrical directional coupler using negatively anisotropic medium

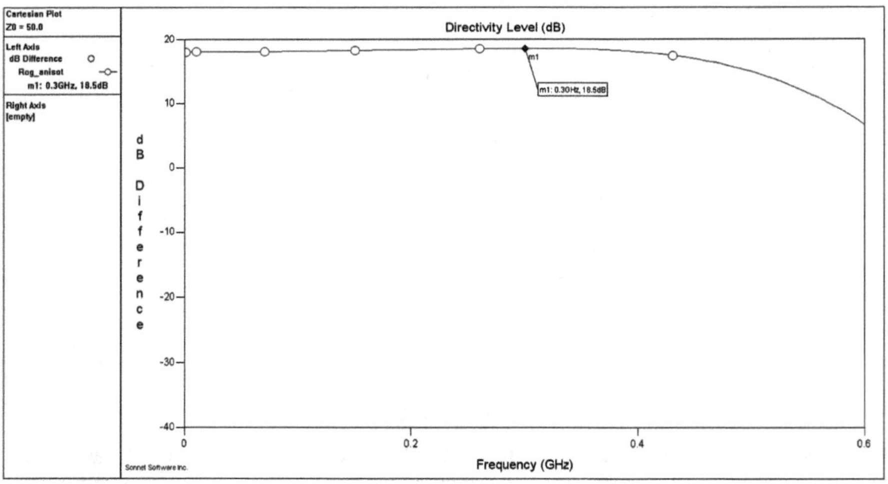

Fig. 8.12 Directivity level for symmetrical directional coupler using negatively anisotropic medium

the equations given in Sect. 8.2.1. The design parameters and physical dimensions using the design (8.92)–(8.123) for 15 dB coupling level at 300 MHz are

$$Material = RO4003 = \varepsilon_r = 3.38$$

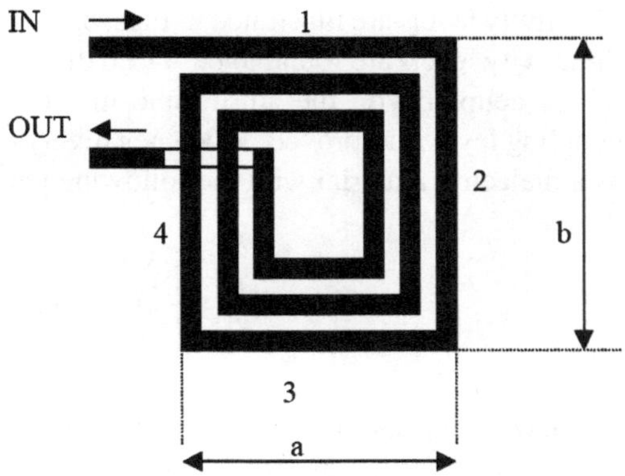

Fig. 8.13 Layout of the spiral inductor for analysis

$$Coupling\ Level = C = -15dB$$

$$f = 300MHz$$

$$\frac{W}{h} = 2.2426$$

$$\frac{s}{h} = 0.3687$$

$$l = 6061.05mils$$

The physical dimensions given above used to simulate the coupler with method of moment based planar electromagnetic simulator, Sonnet. The layout of the coupler is given in Fig. 8.6.

The simulation results showing coupling level and directivity level are illustrated in Figs. 8.7 and 8.8, respectively.

As seen from the figures, the simulated results for the coupling and directivity level at 300 MHz are −14.3 dB, and 11.25 dB, respectively. Now, we keep all the design parameters same and replace the isotropic material, RO4003, with a positively uniaxial anisotropic material with the following permittivity tensor

$$\bar{\bar{\varepsilon}} = \begin{bmatrix} 3.38 & 0 & 0 \\ 0 & 3.38 & 0 \\ 0 & 0 & 5.38 \end{bmatrix}$$

We only change the permittivity of the material along the optic axis to see the effect of anisotropy on the directional coupler response. The coupling and directivity levels are illustrated in Figs. 8.9 and 8.10, respectively. The coupling and directivity levels are found to be −15.6 dB and 3.1 dB, respectively. The directivity of the coupler with the anisotropic medium gets worse drastically whereas the coupling level is improved. When negatively uniaxial anisotropic medium is placed as a dielectric material with the following permittivity tensor

$$\bar{\bar{\varepsilon}} = \begin{bmatrix} 3.38 & 0 & 0 \\ 0 & 3.38 & 0 \\ 0 & 0 & 2.38 \end{bmatrix}$$

the coupling and directivity levels are simulated and found to be −13.46 dB and 18.5 dB, respectively as shown in Figs. 8.11 and 8.12. This is a drastic improvement in the directivity level of the coupler. As a result, it is shown that anisotropy can be used to improve the directivity level and adjust the coupling level of the directional couplers based on the applications.

8.3 Spiral Inductor Design

There is an increasing demand in the area of radio frequency applications to use cost effective planar inductors. Spiral type planar inductors are widely used in the design of power amplifiers, oscillators, microwave switches, combiners, and

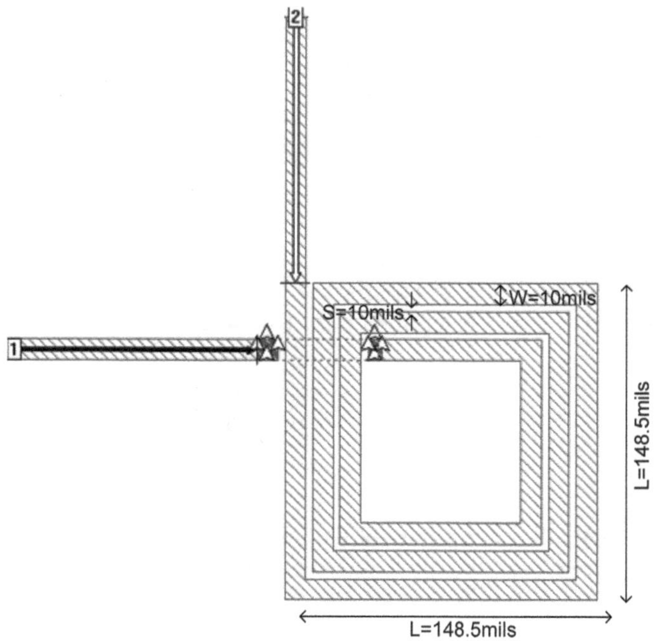

Fig. 8.14 Geometry of the spiral inductor

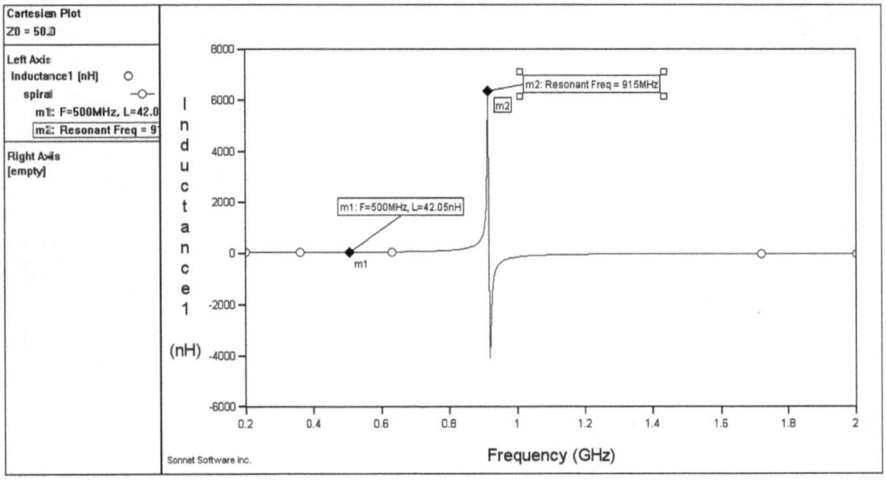

Fig. 8.15 Inductance value of the spiral inductor on isotropic substrate

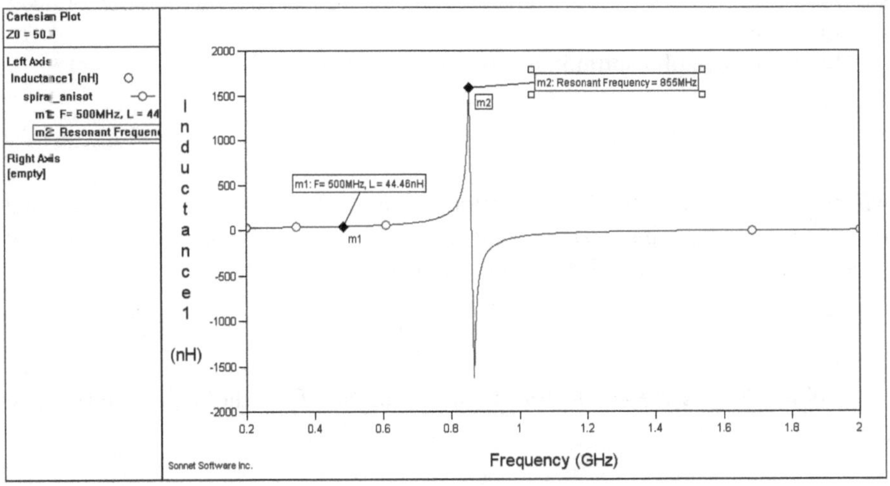

Fig. 8.16 Inductance value of the spiral inductor on an anisotropic substrate

splitters, etc. In this section, we give design examples for the spiral inductors using isotropic media and anisotropic media and study the effect of anisotropy on the inductance value and self resonant frequency. The design equations giving the physical dimensions of spiral type inductors are given in [8] using the quasi-static analysis. The equations take also into account of the mutual inductance that exists in the spiral structure. The layout of the spiral inductor that is used in the analysis is

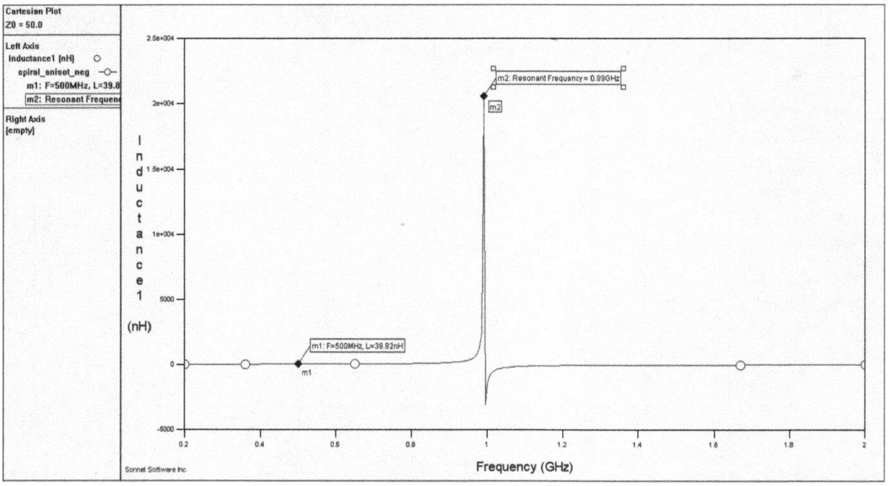

Fig. 8.17 Inductance value of the spiral inductor on a negatively uniaxial anisotropic substrate

shown in Fig. 8.13. When the conductor length a is equal to b, spiral inductor takes the square shape.

The inductance of a spiral inductor is found using

$$L = L_0 + \sum M \tag{8.149}$$

where L_0 represents the self inductance, and $\sum M$ is the sum of all mutual inductances. Mutual inductance is function of length of the conductors and the geometric mean distance, GMD, between them. It is given by

$$M = 2lQ \tag{8.150}$$

M is in nH, l is the conductor length in cm, and Q is the mutual inductance parameter expressed as

$$Q = \ln\left[(l/GMD) + \left[1 + \left(\frac{l^2}{GMD^2}\right)\right]^{1/2}\right] - \left[1 + \left(\frac{GMD^2}{l^2}\right)\right]^{1/2} + \left(\frac{GMD}{l}\right) \tag{8.151}$$

An algorithm using the design equations is written to calculate the physical dimension of a square spiral inductor on an Alumina substrate with $\varepsilon_r = 9.8$ to obtain approximately 40 nH inductance at 500 MHz. The following physical dimension are obtained with the algorithm using the method proposed in [8].

$$\text{Conductor spacing} = s = 3.3 mils$$

$$\text{Width of the trace} = W = 10 mils$$

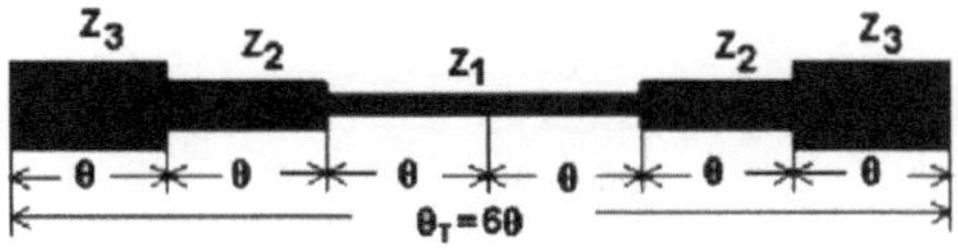

Fig. 8.18 Three section SIR

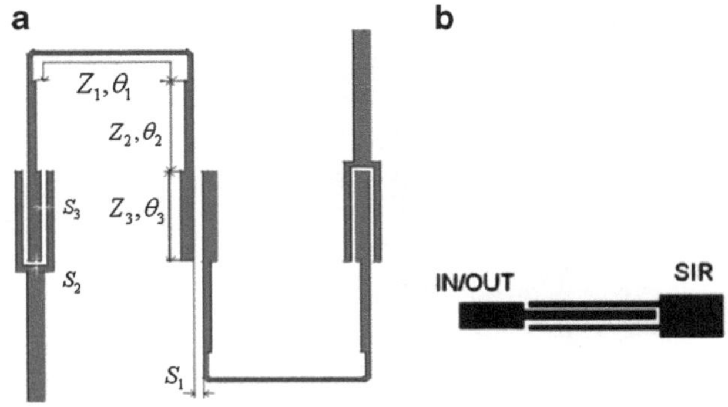

Fig. 8.19 (**a**) Triple band bandpass filter using SIRs. (**b**) Inter-coupled segments

$$\text{Material Type} = \text{Alumina} = \varepsilon_r = 9.8$$

$$\text{Substrate Thickness} = 25mils$$

$$\mathrm{a} = b = 148.5mils$$

$$L = 42nH$$

The total inductance value for the square spiral illustrated in Fig. 8.13 is calculated using the design equations in [8] as $L = 42nH$. Physical dimensions obtained with the quasi-static analysis are used to simulate the spiral inductor with Sonnet electromagnetic simulator.

The geometry used in Sonnet is illustrated in Fig. 8.14. The inductance value of the spiral inductor at 500 MHz is found to be 42.05 nH as illustrated in Fig. 8.15. The resonance occurs at 915 MHz. Now, the material is replaced with a positively uniaxial anisotropic medium with the following permittivity tensor

$$\bar{\bar{\varepsilon}} = \begin{bmatrix} 9.8 & 0 & 0 \\ 0 & 9.8 & 0 \\ 0 & 0 & 11.8 \end{bmatrix}$$

We only change the permittivity of the material along the optic axis to see the effect of anisotropy on the spiral inductor response. The graph showing the inductance value and self resonant frequency is given in Fig. 8.16.

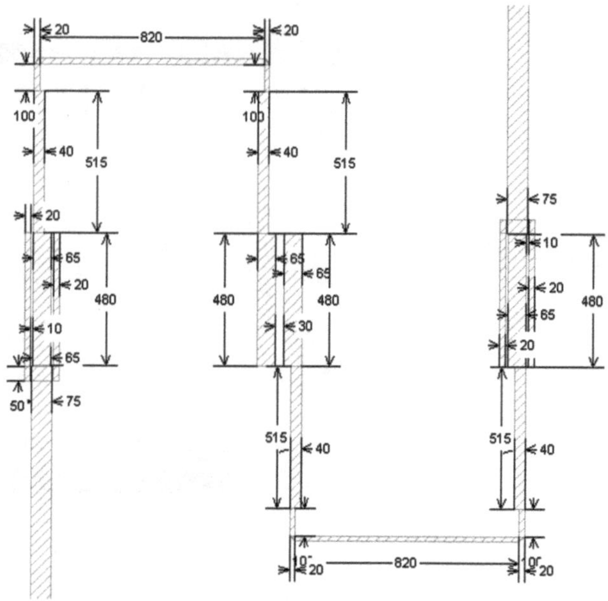

Fig. 8.20 Layout of the triple band bandpass filter simulated by Sonnet

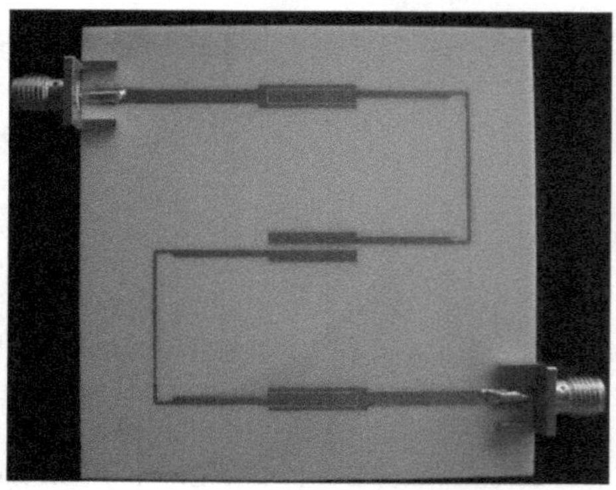

Fig. 8.21 The constructed triple band tri-section bandpass filter using SIRs

It is seen that the change on inductance value is increased by 4.7% whereas the self resonant frequency of the inductor using positively uniaxial anisotropic medium is reduced by 6.56%. When the anisotropic medium becomes negatively uniaxial anisotropic with the following permittivity tensor

$$\bar{\bar{\varepsilon}} = \begin{bmatrix} 9.8 & 0 & 0 \\ 0 & 9.8 & 0 \\ 0 & 0 & 7.8 \end{bmatrix}$$

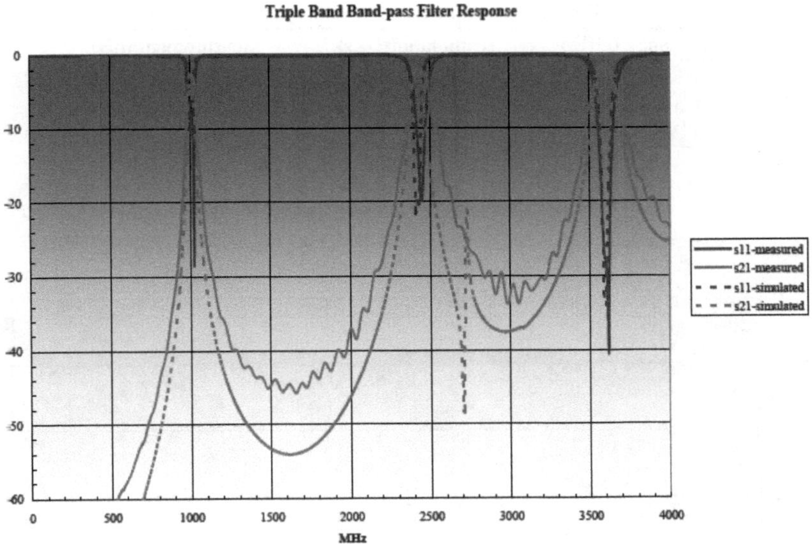

Fig. 8.22 Measured and simulation results for insertion loss and return loss up to 4 GHz

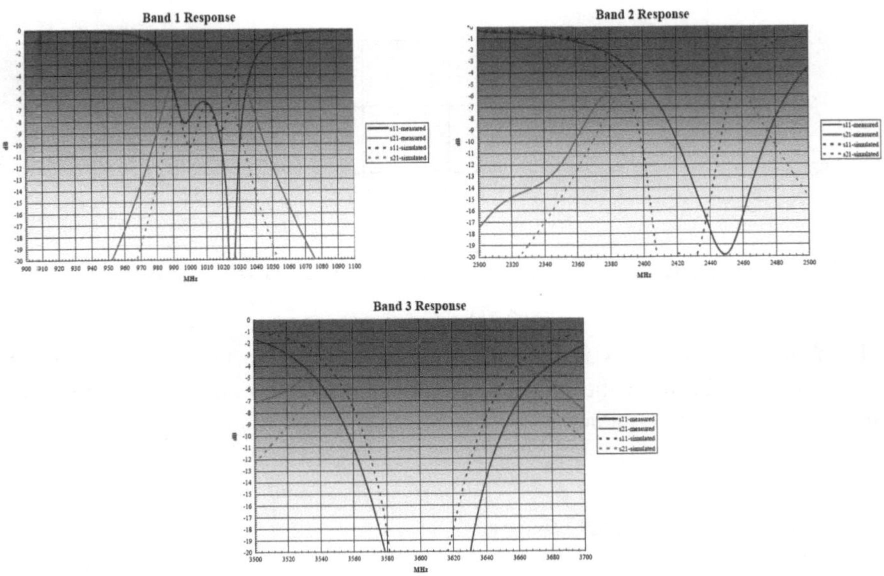

Fig. 8.23 Filter performance in each frequency band

Table 8.10 Tabulation of measurement and simulation results

Frequency (GHz)	Simulation (dB)	Measurement (dB)
1	4	3
2.4	3	4.5
3.6	2	2

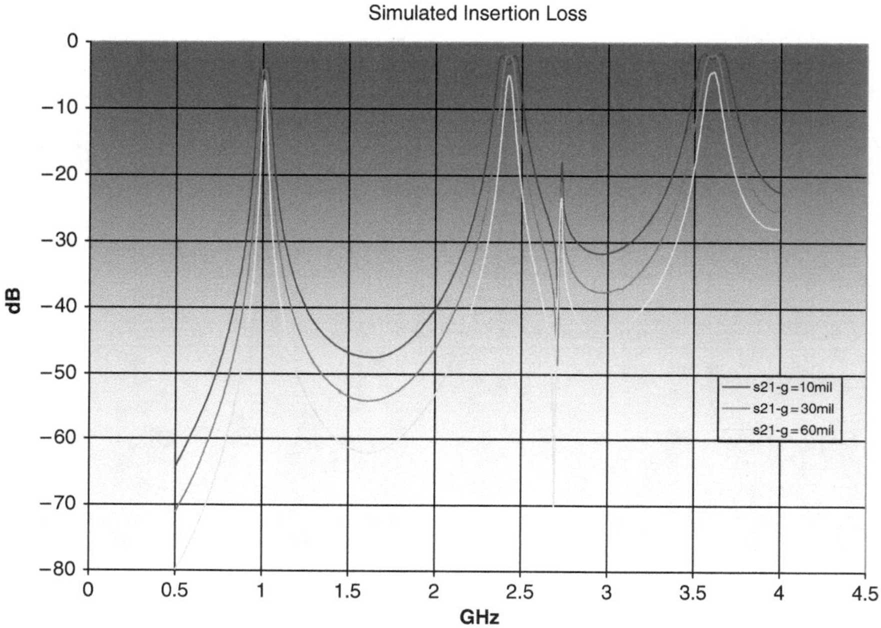

Fig. 8.24 Coupling effect between SIRs on insertion loss up to 4 GHz

this characteristic reverses as expected. This is illustrated in Fig. 8.17. As a result, anisotropy can be used the change the inductance value and the self resonant frequency of the spiral inductor.

8.4 Microstrip Filter Design

Microstrip type filters are one of the indispensable components in RF systems due to their several advantages. Some of these advantages include low cost, manufacturability, and repeatability. In this section, we give design example for a microstrip filters by designing a microstrip triple band bandpass filter using isotropic and anisotropic material. The design procedure will be developed for an isotropic substrate and then will be adapted to anisotropic substrate.

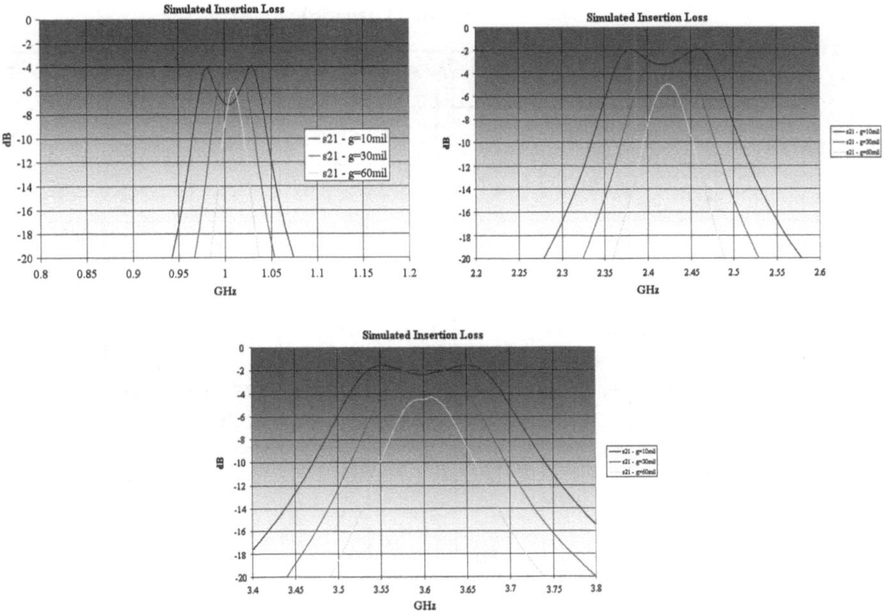

Fig. 8.25 Effect of coupling in each frequency band for tri section triple band bandpass filter

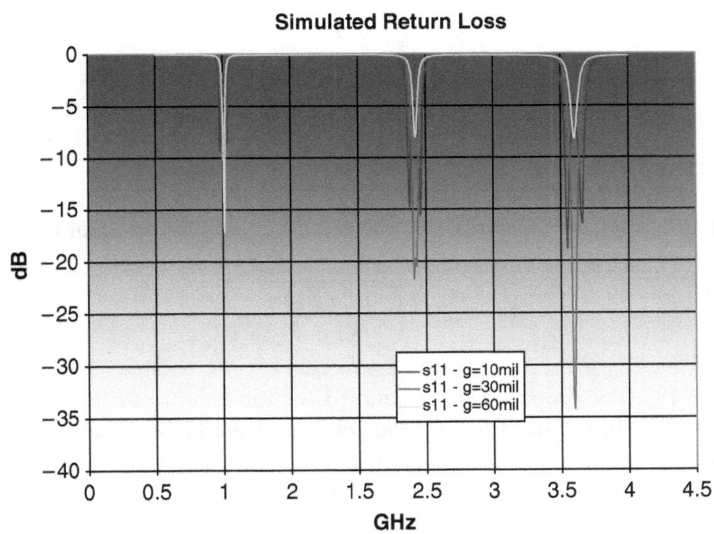

Fig. 8.26 Coupling effect between SIRs for return loss up to 4 GHz

The symmetrical tri-section SIR used in the bandpass filter design is shown in Fig. 8.18.

The total electrical length of the SIR is $\theta_T = 60°$ with impedance ratios

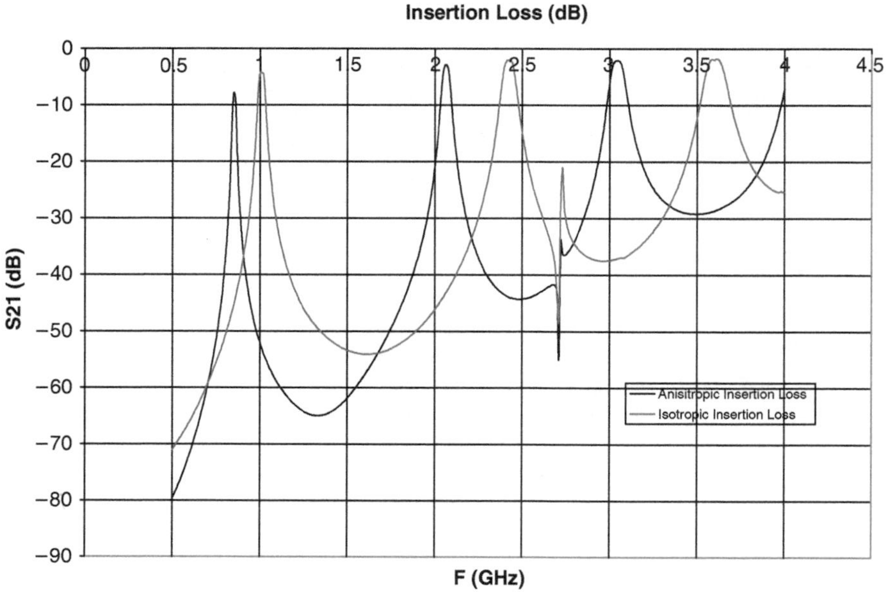

Fig. 8.27 Insertion loss comparison of the filter for isotropic and anisotropic substrates

$$K_1 = \frac{Z_3}{Z_2} \tag{8.152}$$

$$K_2 = \frac{Z_2}{Z_1} \tag{8.153}$$

Hence, the resonance occurs when the electrical length θ is equal to

$$\theta = \tan^{-1}\left(\sqrt{\frac{K_1 K_2}{K_1 + K_2 + 1}}\right) \tag{8.154}$$

as given in [9]. The proposed triple band tri-section bandpass filter using SIRs is shown in Fig. 8.19a below. The wideband characteristic in the passband is obtained by using coupling segments illustrated in Fig 8.19b. Each coupling segment in Fig. 8.19a is capable of producing additional reflection zero (pole) within the band. As a result, inter-coupled open end stub segments in Fig. 8.19b are used to improve the passband characteristics of the filter shown in Fig. 8.19a.

The filter that will be designed has three center frequencies which are located at 1 GHz, 2.4 GHz, and 3.6 GHz. The insertion loss in the passbands is required to be −3 dB or better. The return loss in the first and the second bands is desired to be −20 dB or lower. The third band stopband attenuation is specified to −30 dB or

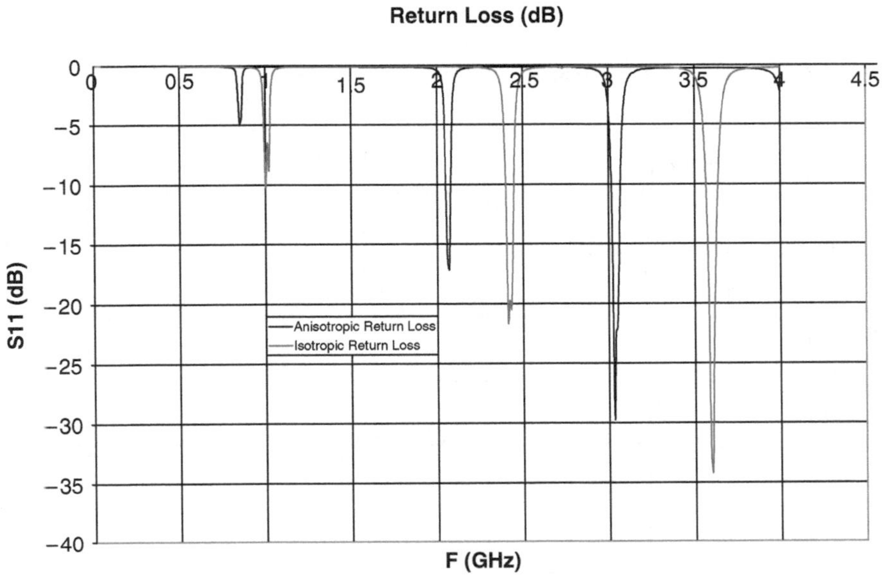

Fig. 8.28 Return loss comparison of the filter for isotropic and anisotropic substrates

lower. The filter topology that will be used is chosen to be second order Chebychev filter topology with 0.1 dB equal ripple in the passband. The dielectric material is chosen to be RO4003 which has a relative dielectric constant of 3.38 and loss tangent of 0.0021. The thickness of the material is 32mil. The physical dimensions of the filter are found using an algorithm developed. The dimensions are then substituted into electromagnetic simulator Sonnet V12 for simulation. The layout of the structure that is simulated in Sonnet is illustrated in Fig. 8.20.

The filter in Fig. 8.20 is built using the dimensions shown on the figure. The final version of the filter that is constructed is demonstrated in Fig. 8.21. The simulation and measurement results for insertion loss, $|S_{21}|$, and return loss, $|S_{11}|$, showing overall performance of the filter up to 4 GHz is illustrated in Fig. 8.22. Fig. 8.23 gives the closer look to the filter performance in each frequency band. Measured and simulated results are found to be closely in agreement. The most deviation between the simulated and measured results in the passband is observed in the second band at 2.4 GHz. The simulated and measured results for insertion loss and return loss are tabulated in Table 8.10. The effect of coupling between each resonator is studied using planar electromagnetic simulator for three different cases in each frequency band. These cases represent different coupling distances which are designated by g between SIRs. The coupling distance, g, is set to be 10mil, 30mil, and 60mil.

The layout shown in Fig. 8.20 is simulated by varying g for the values specified before. The simulation results up to 4 GHz are shown in Fig. 8.24. Figure 8.25 gives closer look to see the effect of coupling in each frequency band for the insertion

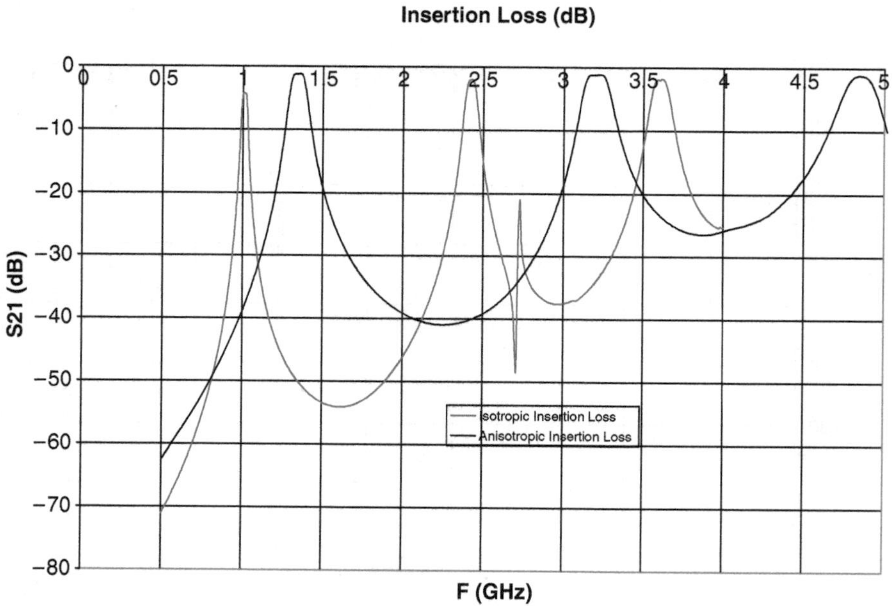

Fig. 8.29 Insertion loss comparison of the filter for isotropic and negatively uniaxial anisotropic substrates

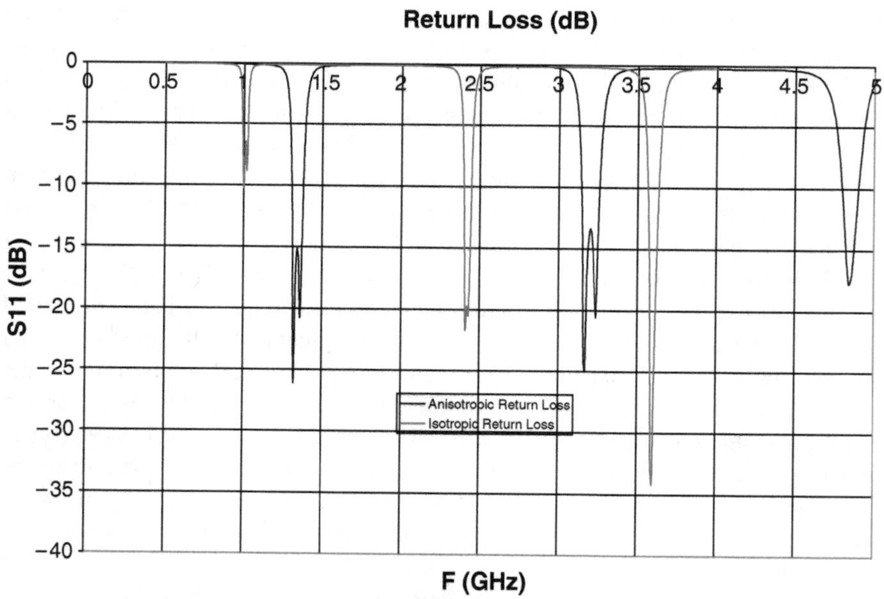

Fig. 8.30 Return loss comparison of the filter for isotropic and negatively uniaxial anisotropic substrates

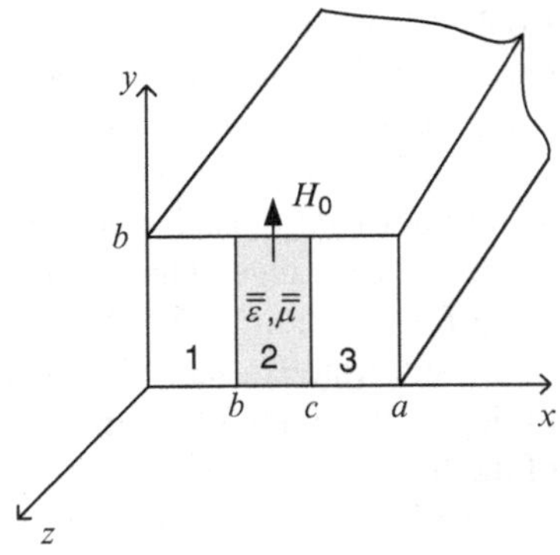

Fig. 8.31 Nonreciprocal phase shifter

loss. It has been observed that as the coupling distance between SIRs is increased, wider bandwidth is obtained for each frequency band in the passband. The return loss of the filter in the stopband for each frequency band when the coupling distance is changed from 10 mils to 60 mils shows different characteristics. In the second and third frequency bands, the return loss of the filter is improved as the gap distance decreases whereas this phenomenon reverses in the first frequency band.

Figure 8.26 shows the return loss response of the filter versus gap distances up to 4 GHz. Based on the results, it is observed that the coupling distance between each SIR can be used as a knob to achieve the desired bandwidth in the passband and attenuation in the stopband for the filter topology used. This feature can be enhanced further if the coupling segments of the SIRs are utilized. For instance, adding a non-uniformity to the coupled lines can be used to set the bandwidth and attenuation only in the desired frequency band without changing the response in the other frequency bands. This is a unique feature for the filters that operate in multiple frequency bands in wireless communication systems. At this point, we replace the isotropic substrate with a positively uniaxial anisotropic medium with the following permittivity tensor

$$\bar{\bar{\varepsilon}} = \begin{bmatrix} 3.38 & 0 & 0 \\ 0 & 3.38 & 0 \\ 0 & 0 & 5.38 \end{bmatrix}$$

As seen from the permittivity tensor structure, we only change the permittivity of the material along the optic axis to see the effect of anisotropy on the filter response. The filter response comparing the insertion loss of the bandpass filter for isotropic and isotropic material is illustrated in Fig. 8.27.

It should be noted that the filter characteristics using anisotropic material as substrate are maintained. However, there is a constant frequency shift in the insertion loss across the bandwidth when an anisotropic substrate is used. The range of the frequency shifts are 0.27 GHz, 0.37 GHz, 0.56 GHz, for the first, second, and third bands respectively. One of the advantages of using anisotropic material in the filter design as a substrate exposes itself in the stopband. In the stopband, it gives better than 10 dB improvement for all frequency bands in comparison to the filter response using isotropic substrate. The filter response comparing the return loss of the bandpass filter for isotropic and anisotropic material is given in Fig. 8.28. The return loss using with anisotropic substrate is again improved in each frequency band around 10 dB similar to the insertion loss response. The range of the frequency shift that exists in the insertion loss response is same for the return loss. This effect is reversed when negatively uniaxial anisotropic substrate is used. The filter response confirming this characteristic is shown in Figs. 8.29 and 8.30 for insertion loss and return loss, respectively. The response of the filter in the stopband also reverses for return loss and insertion loss when the anisotropic substrate is negatively uniaxial accordingly as illustrated.

8.5 Nonreciprocal Phase Shifter Design

Nonreciprocal phase shifters such as ferrite phase shifters have excellent electrical performances and are commonly used as phasing elements in phased array antennas and in many other RF/microwave systems due to their advantages of high Q value, high power handling capability etc. The geometry of nonreciprocal phase shifter using Ferrite as a thin slab in a rectangular waveguide is illustrated in Fig. 8.31.

TE_{m0} modes are derived in Sect. 8.1.2 for a rectangular waveguide filled with gyrotropic medium. For this problem, we simplify the material to be magnetically gyrotropic such as ferrite. This material is placed in region 2 of the rectangular waveguide shown in Fig. 8.31. Regions 1, and 3 in the waveguide are assumed to be filled with air. Then, following the same procedure described in Sects. 8.1.1 and 8.1.2, the field components for TE_{10} modes can be obtained. In the regions filled with air, Region 1, and 3, the electric fields are

$$E_{y1} = C\sin(k_a x) \tag{8.155}$$

$$E_{y3} = D\sin(k_a(a-x)) \tag{8.156}$$

The dispersion relation in these regions is defined as

$$k_a^2 = \omega^2\mu_0\varepsilon_0 - k^2 \tag{8.157}$$

The magnetic field components are found following the procedure outlined before and given as

$$H_{z1} = \frac{-iC}{\omega\mu_3}\cos(k_a x) \tag{8.158}$$

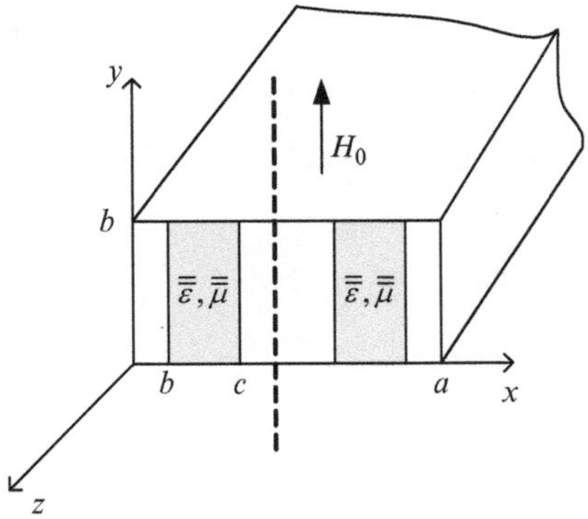

Fig. 8.32 Two slab nonreciprocal phase shifter

$$H_{z2} = \frac{-\mu_1}{\omega(\mu_1^2 - \mu_2^2)} \times \begin{bmatrix} Ak_\perp[\cos(k_\perp x) - i\sin(k_\perp x)] - Bk_\perp[\cos(k_\perp x) + i\sin(k_\perp x)] + \\ -\frac{ik\mu_2}{\mu_1}[\cos(k_\perp x) - i\sin(k_\perp x)] - Bk_\perp[\cos(k_\perp x) + i\sin(k_\perp x)] \end{bmatrix} \tag{8.159}$$

$$H_{z3} = \frac{iDk_a}{\omega\mu_0}\cos(k_a(a-x)) \tag{8.160}$$

$$H_{z1} = \frac{-Ck}{\omega\mu_0}\sin(k_a x) \tag{8.161}$$

$$H_{z2} = -\frac{\mu_1}{\omega(\mu_1^2 - \mu_2^2)} \times \begin{bmatrix} -k_\perp[A[\cos(k_\perp x) - i\sin(k_\perp x)] - B[\cos(k_\perp x) + i\sin(k_\perp x)]] + \\ -\frac{ik\mu_2}{\mu_1}[A[\cos(k_\perp x) - i\sin(k_\perp x)] + B[\cos(k_\perp x) + i\sin(k_\perp x)]] \end{bmatrix} \tag{8.162}$$

$$H_{z3} = \frac{iDk_a}{\omega\mu_0}\cos(k_a(a-x)) \tag{8.163}$$

$$H_{x2} = \frac{\mu_1}{\omega(\mu_1^2 - \mu_2^2)} \times \begin{bmatrix} -k[A[\cos(k_\perp x) - i\sin(k_\perp x)] + B[\cos(k_\perp x) + i\sin(k_\perp x)]] - \\ \frac{ik_\perp\mu_2}{\mu_1}[A[\cos(k_\perp x) - i\sin(k_\perp x)] - B[\cos(k_\perp x) + i\sin(k_\perp x)]] \end{bmatrix} \tag{8.164}$$

$$H_{x3} = -\frac{Dk}{\omega\mu_0}\sin(k_a(a-x)) \tag{8.165}$$

Using the boundary conditions on the walls of the dielectric in the waveguide, we can write

$$E_{y1} = E_{y2},\ x = b \tag{8.166}$$

$$E_{y3} = E_{y2},\ x = c \tag{8.167}$$

Application of the boundary conditions given by (8.166) and (8.67) on the fields given by (8.67), (8.155), and (8.156) leads to following equation

$$\frac{k_a k_\perp \mu_1}{\mu_0(\mu_1^2-\mu_2^2)}[\tan(k_a b) + \tan k_a(a-c)]\cot(k_\perp(b-c)) - \\ + \frac{k_a \mu_2 k}{\mu_0(\mu_1^2-\mu_2^2)}[\tan(k_a b) - \tan k_a(a-c)] - \\ \left[\left(\frac{k_\perp \mu_1}{(\mu_1^2-\mu_2^2)}\right)^2 + \left(\frac{k\mu_2}{(\mu_1^2-\mu_2^2)}\right)^2\right]\tan(k_a b)\tan(k_a(a-c)) + \left(\frac{k_a}{\mu_0}\right)^2 = 0 \tag{8.157}$$

The propagation constant or phase constant is found by solving the equation given in (8.165) with an algorithm developed. The phase shift effect can be extended to the symmetrical structure shown in Fig. 8.32. The analysis is similar to what is presented for the single slab phase shifter.

References

1. C. T., Tai, *Dyadic Green's Functions in Electromagnetic Theory*, 2nd ed., IEEE Press, Piscataway, NJ, 1994.
2. A. Eroglu, "The Complete Design of Microstrip Directional Couplers Using the Synthesis Technique," *IEEE Transactions on Instrumentation and Measurement*, vol. 57, issue 2, pp. 2756–2761, 2008.
3. I. Bahl, *Lumped Elements for RF and Microwave Circuits*. Artech House, Norwood, MA, 2003.
4. M. Kobayadshi, "New view on an anisotropic medium and its application to transformation from anisotropic to isotropic problems," *IEEE Transactions on Microwave Theory and Techniques*, vol. 27, issue 9, pp.769–775, 1979.
5. B. Janiczak, and M. Kitlinski, "Analysis of coupled asymmetric microstrip lines on a ferrite substrate," *Electronic Letters*, vol. 19, issue 19, pp. 779, 781, 1983.
6. M. Geshiro, and T. Itoh, "Analysis of a Coupled Slotline on a Double-Layered Substrate Containing a Magnetized Ferrite," *IEEE Transactions on Microwave Theory and Techniques*, vol. 40, issue 4, pp. 765–768, 1992.
7. M. R. Albuquerque, A.G. D'Assuncao and A. J. Giarola, "On the properties of microstrip directional couplers on magnetized ferrimagnetic layers," *25th European Microwave Conference*, pp. 713–716, 1995.

8. H. M. Greenhouse, “Design of Planar Rectangular Microelectronic Inductors”, *IEEE Transactions on Parts, Hybrids, and Packaging*, vol. PHP-10, No. 2, pp. 101–109, 1974.
9. X. Lin and Q. Chu, “Design of triple-band bandpass Filter using tri-section stepped-impedance resonators,” *International Conference of Microwave and Millimeter Wave Technology*, D1.6, April 2007.

Index

FSC
www.fsc.org
MIX
Papier | Fördert gute Waldnutzung
FSC® C083411

Kolibri 360 GmbH
Ferdinand-Jühlke-Straße 7
99095 Erfurt, Deutschland
produktsicherheit@kolibri360.de